D0998414

STRAND PRICE

PRAISE FOR GREEN FIG AND LIONFISH

"Allen Susser's *Green Fig and Lionfish* (Books & Books Press) takes what we have learned from overfishing and uses it to help the ecosystem, making it truly a book for our times. The idea and the recipes are not to be missed."

—**Mark Kurlansky,** *New York Times* bestselling and James Beard award-winning author of *Salt: A World History, Cod: A Biography of the Fish that Changed the World,* and *Milk!: A 10,000-Year Food Fracas*

"From Florida, throughout the Caribbean, and down to my home of Saint Lucia, Chef Allen Susser raises awareness of the real threat posed by lionfish and a yummy way to help eliminate it. Fortunately, the pesky invader is quite tasty, and Chef Allen's fun recipes will brighten your table with their bold island flavors. Let's get them off the reef and onto your plate!"

—**Nina Compton,** Saint Lucian culinary ambassador and native; chef/owner of Compère Lapin in New Orleans

"At Le Bernardin, we spend our days evaluating seafood, which means not simply the quality, but also the ethics and sustainability of how they are harvested. We believe deeply in supporting the artisanal fishers who are seeking out the most sustainable methods and species. What we've recently discovered is that lionfish, an extremely invasive and detrimental species, are not difficult to catch and in fact have delicious and very versatile flesh. They are truly one of the best examples of sustainable fishing, which I'm happy to encourage!"

—**Éric Ripert,** author of *32 Yolks: From the Mother's Table to Working the Line* and chef/CEO-owner of Le Bernardin restaurant in New York

"I love the ocean and I love to eat sustainable seafood. We must continue to keep our oceans healthy for future generations and for our own survival."

—**Cindy Pawlcyn,** author of *Mustards Grill Napa Valley*, chef/owner of Mustard's Grill in Napa Valley, and James Beard award-winning chef

"Chef Allen has always been a champion of local, seasonal cooking, even before it was 'cool.' With this cookbook, he is promoting something even cooler...the sustainability of the seas, one of our most important natural resources, by giving us many amazing recipes for the damaging (and delicious!) lionfish."

—**José Andrés,** author of several books, including *Made in Spain*, chef/owner of ThinkFoodGroup, educator, television personality, and humanitarian

"The first time I saw a lionfish in the wild was many years ago while snorkeling off the beach on Andros Island, Bahamas. Since then, working as a fisherman and diver from Bimini to the Keys and beyond, I have seen firsthand the devastation that these predators can cause to our fragile reef system. They're easy to spear and have no fear of divers, and their stings are VERY plentiful (unfortunately), but the good news is they are out-of-this-world table fare. Spearing them, we have eaten well. I only wish my dear friend Chef Allen was with us to cook! Chef Allen loves South Florida, its people, and our natural resources. This is the perfect time for this book, and he's the perfect person to write it!"

—**Paul Castronovo,** Radio personality of BIG 105.9, fisherman, diver, and member of the Board of Directors of both the Guy Harvey Ocean Foundation and OCEARCH

"Making the right choices about what we eat has a broad impact on the world around us. In the case of lionfish, choosing to eat it as often as possible is the best way to protect our oceans. Eat them to beat them!"

—**Sheila Bowman,** Seafood Watch Manager of Culinary and Strategic Initiatives at the Monterey Bay Aquarium, whose mission is to inspire conservation of the ocean

"I've known Allen Susser and followed his career with avid interest ever since back in the 1980s, when he and some young colleagues pioneered what was then called 'Floribbean Cuisine,' and his new cookbook brings into twenty-first century focus everything he has learned and created in a style now uniquely his own, one highly influential among his peers both in Florida and throughout the Caribbean."

—**John Mariani,** one of America's premier food writers with multiple James Beard Awards, lead food and travel correspondent for *Esquire* magazine, and author of several of the most highly regarded books on food, including *The Encyclopedia of American Food & Drink*

GREEN FIG AND LIONFISH

GREEN FIG AND LIONFISH

Sustainable Caribbean Cooking

By ALLEN SUSSER

BOOKS & BOOKS PRESS
CORAL GABLES

Published by Books & Books Press, an imprint of Mango Publishing, a division of Mango Media Inc.

Cover, Layout & Design: Morgane Leoni
All photo credits: © Bernd Rac

For permission requests, please contact the publisher at:

Books & Books Press
Mango Publishing Group
2850 S Douglas Road, 2nd Floor
Coral Gables, FL 33134 USA
info@mango.bz

For special orders, quantity sales, course adoptions and corporate sales, please email the publisher at sales@mango.bz. For trade and wholesale sales, please contact Ingram Publisher Services at: customer.service@ingramcontent.com or +1.800.509.4887.

Green Fig and Lionfish: Sustainable Caribbean Cooking

Library of Congress Cataloging-in-Publication number: 2019948621
ISBN: (print) 978-1-64250-164-3, (ebook) 978-1-64250-165-0
BISAC: CKB076000—COOKING / Specific Ingredients / Seafood

Printed in the United States of America.

Dedicated to the generations of caring women in my life: my grandmother Rose, mother Molly, wife Judi, and my girls Deanna and Liza, who helped me to understand how important it is for us all to come together at the table.

TABLE OF CONTENTS

INTRODUCTION

The story of the lionfish, like all good fish tales, is growing greater each day. These flamboyantly colorful fish with their diverse markings were originally bred in the Indian Ocean. But legend has it that they found their way over to us in the Caribbean when Hurricane Andrew liberated a handful of them from the aquariums of drug lords living in Florida.

Since then, lionfish have bred prolifically in a ceaseless invasion of our seas from Florida to the West Indies. They thrive in the warm waters of the Atlantic and Caribbean and wreak havoc on our ocean ecosystems and fisheries, gobbling up reef fish, juvenile snapper, and grouper. With no known predators to stop them, the lionfish are more threatening than they are beautiful. Not only are they dangerous to fragile ecosystems, they can also inflict an extremely painful sting on humans. Therefore, we need to jump in and put these delicious fish on our dinner plates.

My first encounter with lionfish was at Anse Chastanet in Saint Lucia, where I am the consulting chef for both Anse Chastanet Resort and Jade Mountain. One of the local divers brought in a handful of lionfish from his morning's catch. We purchase all our fish from local fisherman and usually get our lobster from the divers. This diver was complaining about how these little monsters were eating everything on the reef and destroying the lobster grounds. Would we buy lionfish from him, too?

I had heard of lionfish and was even aware that they were spreading rapidly into the Caribbean. I understood the problems they were causing and, at the same time, was very curious to taste this little creature. And here they were, six of them dangling in front of me, practically dripping with seawater. My chef's inquisitiveness had gotten the better of me: love at first bite. I was encouraged and enthusiastic. The experience stoked my quest for delicious tasting local sustainable ingredients. And that quest has evolved into this cookbook.

One challenge: catching these things isn't so easy. Lionfish don't go for baited hooks, don't school together, and are well camouflaged against the reefs they populate. Divers spear or net each fish, bringing ashore their bounty in a celebration of sustainability. While lionfish haven't often appeared on menus in the Caribbean, more restaurants and fishmongers on many islands are beginning to introduce the fish as well as its story as the fish becomes more prevalent in these waters. In Saint

Lucia, lionfish are becoming a local commodity. You can find them early mornings in the fish market in town.

You will find this seasonal lionfish cookbook a practical guide to cooking fish. Lionfish fillets are dense, with an elastic bite; a sweet, briny flavor; and just a hint of buttery richness. I like to use lionfish in dishes such as ceviche or simply sautéed with cilantro, garlic, and lime. Delicious grilled, sautéed, or stewed, it's a near perfect stand-in for many recipes likely in your repertoire already. The key to fish cookery is keeping it fresh and simple. The fresher the fish, the brighter and more honest the flavor. My instinct is to let flavor come through clearly with these simple recipes.

With all due respect for the people of Saint Lucia, I have borrowed the name of this book, *Green Fig and Lionfish*, from the essence of their namesake, the national dish Green Fig and Saltfish. True to the history of Saint Lucia, mature green bananas are called green figs and play a vital role in daily local cooking. Taking a leap here, I have swapped lionfish for the saltfish that is usually in the pot. I endeavor to not alter the character of the ingredients but to enable them to embody the soulful essence of the Caribbean.

CARIBBEAN CULINARY ESSENTIALS

Seasonal Caribbean

Most people think of the Caribbean as being synonymous with year-round summer. They do not think of it as a place that has seasons, but it does; it has a rhythm of its own. There is a dry season that links winter to spring and a wet season that carries summer through to the fall.

Seasons in the Caribbean influence daily life. That is because most of the food eaten day-to-day by locals is grown on an island. As a chef, I want the best and tastiest ingredients. When cooking with what is in season, you are using ingredients with a higher nutritional value and optimal flavor. Mr. Troubetzkoy, the owner and architect of Jade Mountain, understood what it meant to have local and unique organic Caribbean ingredients for his guests to enjoy. He established his Emerald Estate, a thirty-acre organic farm, to be our source of high quality custom grown fresh tropical fruits, greens, vegetables, and spices solely for the resort's restaurants.

Taste over trend: Caribbean seasonal flavors dance to a tropical beat. It's almost impossible to talk about inspired cooking without talking about seasonality. Choosing and working with the right ingredients isn't hard, but it does require a little know-how and planning to select at optimum taste.

If you spend time in Saint Lucia as I have, from year to year you can experience the seasons' natural rhythms blending into each other. My favorite is mango season, which begins with beautiful purple or yellow blossoms in the dry season around January. The treetops of most villages blush with these floral colors. Within weeks, tiny pea sized mangos set, replacing the fading flowers. As the rainy season begins, you can see the transformation as the tree branches droop, heavy with clusters of green mangos. And by mid-rainy season, between June and July, the island is dripping with lush mangos in all colors of the rainbow: yellow, red, green, and purple. Mangos are happily devoured fresh out of one's hand, chopped into salsa and chow chow, or cooked into jams, chutneys, curries, and pepper pots.

In Saint Lucia, the winter markets are filled with pineapple, carambola, and passion fruit along with locally grown citrus, including limes, shaddocks, tangerines, sour oranges, and sweet oranges. You may find that oranges here often have a green skin, as if they are not ripe yet. That is because the nights are not cool enough for the natural chemical reaction to turn them orange. Nonetheless, the local oranges are just as refreshingly juicy and sweet. The Soufriere street market is always beaming with red, green, and yellow seasoning peppers, sweet peppers, and hot peppers. The spice vendors hawk their ginger, turmeric, cinnamon, nutmeg, cloves, allspice, chandon beni, cilantro, thyme, scallions, and scotch bonnets.

You can always find someone at the market year-round with freshly harvested coconuts piled high in the back of an old pickup, cracking them with his machete. Taste the coconut's invigorating juice and share some of its sweet tender white flesh. Wooden tabletops are filled with rooty ground provisions such as earthy-tasting yams, sweet potatoes, dasheens, plantains, and green bananas, which are staples in local and Caribbean cooking.

As the tropics heat up into the rainy season, mango, papaya, soursop, watermelon, and guava are abundant. Be on the lookout for breadfruit, chayote, and avocado as they reach their peak season. If you are walking through the market, you are going to want to grab some of these treasures. There is nothing like a tree-ripened, sun-kissed fruit grown just a few miles down the road that was probably picked just that morning.

Happily, seasonal cooking is more sustainable. By focusing on locally grown food, you don't have to navigate seasonality—that just falls into place naturally. At home in your kitchen, use the seasons to your advantage with beautiful, colorful, and fresh produce as well as fish and shellfish. Feel free to be creative in using these recipes for Caribbean inspiration to adapt to your local regional ingredients.

Fish Cookery

Fish is a backbone of Caribbean life. So much nourishment comes from the sea that surrounds these islands. This simple guide can be as practical in fish cookery of snapper, grouper, mahi-mahi, or kingfish as it is with lionfish. The key to selecting delicious, top quality fish is using your nose. Fresh fish has a clean aroma, a little like the ocean or a salty breeze from a tropical beach.

Starting off with fresh, pristine fish is essential. Have the fishmonger skin the lionfish and remove all the venomous spines so you do not need to worry about any of this. Give your fish a good rinse under cold running water before you cook it. Many home cooks in the Caribbean often add a squeeze of fresh lime juice at this point too.

Baking

Baking is simple and mess free. You can keep your oven temperature moderate and cook the fish gently. I suggest using a pan or ovenproof dish that is just larger than the fish you are baking to catch the flavorful juices that are released close to the fish.

Stewing

Anything and everything is stewed throughout the Caribbean. One-pot cooking has its island traditions based on necessities of conservation and survival. Pot fish, usually a variety of different small fish including lionfish, often find their way into a stove top stew of spices, herbs, peppers, tomatoes, onions, and broth.

Braising

Braising is a common practice in Caribbean fish cookery. Simply place the fish in a pot full of fragrant spices, stock, and vegetables. Put it in the oven and the fish is cooked in those aromatic juices.

Poaching

Poaching is a classic way to cook fish and one of the easiest. It treats fish very gently and results in a tender, flaky, and moist fish. The intensity of the broth surely can make a difference. In the Caribbean, this can range greatly depending on the island's culinary influences from the French, Spanish, English, Portuguese, or Dutch and how it is balanced by local Creole flavor.

Grilling

Charcoal and wood fire grilling will impart maximum flavor. Build a good fire and let it burn down to hot white ash embers. Clean and oil the grates and liberally brush the fish with oil or butter. When you put the fish on the grill, do not move it around. After a few minutes, just flip it once and, in another minute, off it goes, keeping the flesh moist.

Deep Frying

Deep frying has been a staple of Caribbean fish cooking for generations, yielding crispy, crunchy whole fish, fillets, and chips. This method has also been the most abused and misunderstood. Doing it correctly requires clean vegetable oil (peanut, coconut, or canola oil) to be preheated to a temperature of 375 to 400 degrees to start. Being careful not to overcrowd the deep fryer. Frying the fish golden brown and draining well will bring results that should be finger-licking delicious.

Sautéing

Cooking fish in a fry pan with just enough butter or oil is a simple way to get great tasting results. With lionfish, lightly seasoning the fillets and preheating the pan with the cooking fat will allow the fish to brown quickly. Sautéing is done over moderate heat to allow both sides to be cooked through and finish with a golden brown, moist, and flaky fillet.

Pan Roasting

Pan roasting may be a common restaurant technique, but I feel it serves the home cook well. As with sautéing, use an ovenproof pan that is preheated with just enough oil or butter. However, after sautéing the first side of the fish lightly, it is flipped, and the whole pan is placed in a hot, preheated 375-degree oven to finish cooking with a bit more general heat surrounding it to keep it moist. Aromatic vegetables are often added to the pan to marry the flavors.

CHASTANET
JADE MOUNTAIN
Elijah Jules
Chef de Cuisine

Pan Frying

Pan frying is a lot like sautéing but with a lot more oil or butter. The extra fat prevents the fish from sticking while delivering a high-temperature cooking treatment for a crisp exterior.

Spicing the Caribbean

The Caribbean is a giant melting pot of flavors. Caribbean cuisine is influenced by the cooking of many other cultures including India, Africa, Asia, and Europe. Caribbean food is bold, flavorful, spicy, hot, and fresh, and it is truly a world cuisine that stands on its own. The spices that I discuss are not only used in abundance here in the Caribbean, but they were transplanted from their original habitat two or three centuries ago by the European powers of the day in their bid to control the lucrative world spice trade.

The Caribbean Islands constitute a massive archipelago located in the Caribbean Sea that can be subdivided into a few different regions: the Lucayan Archipelago (the Bahamas and the Turks and Caicos Islands), the Greater Antilles (including the Cayman Islands, Jamaica, Puerto Rico, the Dominican Republic, Cuba, and Haiti), and the Lesser Antilles or West Indies, including the Leeward and Windward Islands (St. Barthélemy a.k.a. St. Barts, Barbados, Saint Lucia, and Trinidad and Tobago) and the ABC Islands (Aruba, Bonaire, and Curaçao). There are thirteen sovereign states and seventeen dependent territories in the Caribbean, and the predominant languages are English, Spanish, French, Dutch, and Antillean Creole. The Caribbean sprawls across more than one million square miles and is primarily located between North and South America.

Just as there are many different islands, there are many herbs and spices used in Caribbean cooking. Saint Lucia has a warm and sunny climate with a year-round growing season. Mother Nature is in charge of doling out a dry season and a wet season. Each offers its own unique tropical fruits, citrus, and vegetables, including tree vegetables and root vegetables, for distinct seasonal flavor. Learn to choose and use each season's best.

With over one hundred spices, leaves, flowers, and herbs used in Caribbean cooking, it is difficult to narrow them down into a simple spice box. The most frequently used herbs include thyme, marjoram, basil, chadon beni, and green onion. Herbs and spices are regularly used for seasoning, and, to this day, they are also still frequently used in local bush remedies that have been passed down from grandmother to mother to daughter. You will probably find some combination of these five spices throughout island cooking.

Caribbean Spices

- **Cinnamon** is also known loosely as "spice" or "hard spice" locally. It has a sweet, fragrant, woody aroma. Its roots can be traced back to biblical days. It was brought to the Caribbean by the English East India Company.
- **Ginger root**, introduced to the West Indies by the Spaniards, is a must-have in every Caribbean kitchen. The fresh root is grated to release its sweet rich undertones to flavor several savory dishes with a slightly biting heat.
- **Cloves** are used in both savory and sweet preparations. One of the top uses of cloves is as an aromatic component in a curry sauce. The Dutch, by the way, held the monopoly on this spice from the 1600s. Cloves can be used whole or ground and have a tasty assertive dark aroma.
- **Nutmeg** is a pantry staple that is used for beverages such as Caribbean rum punch and other alcoholic drinks. The English planted nutmeg heavily in Grenada toward the end of the eighteenth century. Nutmeg is best when freshly grated. Its rich, fresh, and warm aroma is used to flavor sauces and bitter greens.
- **Allspice** is a spice that comes from the dried fruit of the allspice evergreen tree or pimento tree. This is the Caribbean's native spice. It is not a combination of spices as is commonly thought. It is called allspice because when ground, the spice berries taste like a combination of nutmeg, cinnamon, black pepper, and cloves. It was this spice that Columbus brought back to Spain, thinking he had found the source of pepper.

Colorful, juicy, crunchy, sweet, bitter, herby, spicy... One of the great things about the Caribbean is that there are so many flavors and textures to put together, making it exciting to cook. The kitchen is freedom.

Caribbean Spice Box

CREOLE SPICE

- 1 teaspoon ginger
- 2 teaspoons coriander
- 1 teaspoon allspice
- ½ teaspoon cinnamon
- 2 teaspoons cumin
- ½ teaspoon crushed red pepper flakes
- 1 teaspoon kosher salt

JERK SPICE

- 1 teaspoon cinnamon
- 2 teaspoons allspice
- 1 teaspoon clove
- 1 teaspoon nutmeg
- 2 teaspoons black pepper
- 1 teaspoon crushed red pepper flakes
- 1 teaspoon kosher salt

WEST INDIAN SPICE

- 1 teaspoon ginger
- 1 teaspoon coriander
- ½ teaspoon cardamom
- 1 teaspoon black pepper
- ½ teaspoon cinnamon
- 1 teaspoon star anise
- 1 teaspoon turmeric
- 1 teaspoon kosher salt

SUSTAINABILITY

For us at Jade Mountain and Anse Chastanet, sustainability is not a lofty idea but a fundamental and necessary endeavor. Our concept is centered on the food and drink, of course, but it's also about our team, our facilities, our practices, and the hundreds of decisions we make each day that affect the world around us. We believe it's about finding a balance that allows us to sustain our quest of making quality, accessible food while also giving back to our community and the environment.

Throughout coastal Florida as well as the Caribbean, coral reefs are home to a wide variety of plants and animals. Some of these reefs date back over fifty million years. Many of the small reef fish feed on the plants and tiny creatures that make up the reef. In the natural course of the food chain, little fish are nourishment for bigger fish. Larger fish become prey for larger sea creatures, and so on, up to the apex predators of the deep.

Not only are the reefs at risk due to overfishing, pollution, and climate change, but lionfish are devastating many of the reefs in the Caribbean. They are eating many times their fair share of the nourishment present in the habitat. They have no predators in these waters as they are an invasive species. Therefore, we need to step in and eat lionfish.

Sustainability Recommendations

For Red lionfish (Pterois volitans) and Devil firefish (Pterois miles)

These two species are caught by spear and as incidental bycatch in the Florida Caribbean spiny lobster pot and lobster trap fishery. The lionfish fishery is found in the western Atlantic Ocean, the Caribbean Sea, and the Gulf of Mexico. There is low conservation concern, as lionfish are an invasive species outside of the Indo-Pacific and are detrimentally affecting native species through predation and resource competition. As a result, managers are focusing on ways to reduce and prevent further spread of the lionfish population. There are no bycatch species for the spearfish fishery since lionfish are targeted.

There is currently no fishery management plan for lionfish aimed at conserving stock size, but multiple control plans are in the process of being developed among local, state, federal, and international partners and are available. There are no regulations in place for the lionfish fishery in Atlantic or Gulf state waters, but it is illegal to transport and release live lionfish.

Lionfish are fished with spears and traps. These gear types tend to cause moderate to no impact on benthic habitats. Spiny lobster traps are deployed in a variety of habitats where they will not harm the rocky reefs and coral, but often in sand and seagrass areas, so gear impact will vary with habitat. Finally, lionfish are both competitors with and apex predators on ecologically, commercially, and recreationally important species; hence, their reduction or removal from the Atlantic, Caribbean, and Gulf of Mexico will greatly benefit the native species.

Excerpt from Monterey Bay Aquarium Seafood Watch Report

Monterey Bay Aquarium's Seafood Watch program evaluates the ecological sustainability of both wild-caught and farmed seafood commonly found in the United States marketplace. Seafood Watch defines sustainable seafood as originating from sources, whether wild-caught or farmed, which can maintain or increase production in the long term without jeopardizing the structure or function of affected ecosystems. Seafood Watch makes its science-based recommendations available to the public in the form of regional pocket guides that can be downloaded from www.seafoodwatch.org. The program's goals are to raise awareness of important ocean conservation issues and empower seafood consumers and businesses to make choices for healthy oceans.

Turning the Tables

Lionfish are unwelcome and do not belong in the tropical Atlantic. Their natural domain is in the South Pacific and Indian Oceans, where they are a normal part of the reef ecosystem. Here in the Caribbean, they are wreaking havoc on the reefs and coastal waters almost all the way around the Gulf of Mexico and up and down the eastern seaboard of the United States. To counter this destructive trend, many active stakeholders and researchers are doing their part to eradicate the problem. Here, we are trying to encourage the commercial hunting of the species as well as volunteer lionfish culls and derbies. Marine scientists recommend catching and eating lionfish as one of the most effective ways of curbing their population growth and controlling their expansion. When you choose to buy lionfish, you're helping to prevent the spread of this invasive species.

In the Caribbean, it's common to see lionfish hovering above the reefs throughout the day and gathering in groups of up to ten or more on a single coral head. They are ferocious eaters that typically flutter their fins to herd smaller fish into a group, then when they have cornered their prey, they pounce. Divers are using a similar strategy; as the lionfish gather together to eat, it is the opportune time for divers to strike with their spears, when they can spear one after another in a maximally productive way. Hunter divers can capture hundreds in a single day.

With fishing comes fishing tournaments. And over the past few years, several lionfish derbies have launched throughout the region; these are often organized as special day-long or weekend-long tournaments with generous cash prizes for winners. These events are primarily for the purpose of collecting, removing, and, most importantly, eating as many lionfish as possible. They have proven to be incredibly popular among local fisherman, dive groups, boat owners, and their families and friends. Not only can you enjoy tasting lionfish in every form imaginable at these events, you can also relish taking part in the accompanying festivities, which often include live music and cooking demonstrations by chefs. Notably, these derby events help to draw media attention to the Atlantic lionfish invasion as well as help grow the commercial lionfish market. This sustainable awareness is one of the goals of this lionfish book.

Efforts are underway in many coastal regions of Florida to reduce the lionfish population through lionfish rodeos and derbies in which spear fishers dive and spear the animals in their habitat. REEF, the Reef Environmental Education Foundation, has been hosting annual lionfish derbies for over ten years. Currently the REEF Lionfish Series Championships in Florida include an annual winter lionfish derby in Key Largo and summer series in Fort Lauderdale and Sarasota. The Mote Marine Laboratory and Aquarium currently hosts an annual lionfish culinary competition. Additionally, there is an annual Palm Beach County Lionfish Derby and Festival as well as the Annual Upper Keys Lionfish Derby and Festival Series Finale. I have witnessed the mayhem and joy of the competing diver teams coming back in late afternoon, their ice chests brimming with writhing lionfish. Whoever catches the largest, the smallest, and the grand prize fish and the team that lands the most lionfish that day have a chance to claim the derby as well as bragging rights for the entire year.

Solving the lionfish problem has a delicious solution! We need to make the lionfish available to chefs and home cooks alike. A positive step in this direction is happening in Florida at Whole Foods Market. Since the Monterey Bay Aquarium rated lionfish "Green" in 2016, Whole Foods Market offers lionfish at all their Florida stores, creating a market for this tasty predator and hopefully making a dent in the growing lionfish population.

In Grenada, there is an annual dive fest to hunt lionfish. In the Cayman Islands, the Cayman United Lionfish League (CULL) organizes lionfish culling tournaments every quarter, with each hunt culminating in an island-wide lionfish feast. And in Curaçao, divers can join lionfish hunts organized by several dive operators. St. Barts recently held its own lionfish rodeo to engage the island community in the effort to mitigate the invasive species in the St. Barts Nature Reserve.

For small-island states in the Caribbean like Barbados, the oceans are a critical element to the survival of the people who live there, so they are disproportionately affected by issues such as rising sea levels, overfishing, and intruding predatory species like lionfish. Forward-thinking local initiatives focus on promoting sustainable fishing methods and supporting responsible artisanal fisherfolk.

A lionfish derby is held annually in collaboration with the Barbados Lionfish Project. Members of this community are active in Slow Fish, the official Slow Food® campaign for sustainable fisheries. The main objective of Slow Fish's social media activity is to educate people about what they can do to make change through their eating choices and habits.

SCUBAPRO
A NATURAL PREDATOR EMERGES

In the Virgin Islands, the Caribbean Lionfish Response Program developed by CORE has been assisting with Invasive Lionfish Management since 2009. Educational materials and training have been provided for the US Virgin Islands, the British Virgin Islands, and Puerto Rico. Programming has consisted of public awareness, education, educational materials, training of invasive lionfish responders, maintaining a lionfish sighting network online at their website, and implementing a lionfish response network.

The Bahamas Department of Marine Resources team is working to collect data on lionfish populations. Experimental removals of lionfish were conducted to see what it will take to control lionfish populations locally and to determine how lionfish are affecting native fish populations in these habitats. Preliminary results indicate that while it may not be possible to eradicate lionfish in the Bahamas, their abundance can be reduced at specific sites by conducting removals every three months. These results will help marine resource managers determine appropriate actions to take to control populations of this invasive predator. Lionfish is appearing more and more frequently on the menus of Bahamian restaurants throughout the islands and even at local farmers' markets throughout the year. Green Turtle Cay in Abaco hosts an annual lionfish derby.

In Jamaica, a campaign aimed at wiping out the species around the island nation has the motto "Eat sustainable, eat lionfish!" That push is sponsored by the US National Oceanic and Atmospheric Administration (NOAA). The National Environment and Planning Agency in Jamaica announced a 66 percent drop in sightings of the invasive fish as people are eating and enjoying the taste of the lionfish in large numbers.

In the Dominican Republic, the La Caleta Fishermen Cooperative is taking a proactive approach by regularly catching and selling lionfish to local restaurants to support the national strategy, which includes encouraging lionfish to be used on local menus. In Aruba, the Aruba Lionfish Initiative Foundation hosts monthly lionfish hunts around the island and currently maintains a daily lionfish cull count.

Here in Saint Lucia, Anse Chastanet is the only place that provides easy and sustainable access to divers due to its progressive green architecture by minimizing impervious surfaces, keeping shorelines stable, and completely preventing runoff from entering the water. Scuba St Lucia, the island's five-star Professional Association of Diving Instructors (PADI) operation, provides the only beach entry to this incredible experience.

Scuba St Lucia now fuses marine conservation, sport fishing, and fine dining with a new innovative and unique lionfish removal program to fortify their aggressive conservation initiatives. Visitors are now authorized by special permit to spearfish the invasive lionfish as part of a win-win sustainable tourism partnership. Scuba St Lucia has also added a PADI "Invasive Lionfish Tracker Specialty Course" which provides two dives to educate visitors about humanely controlling the invasive species.

Taking angling to the next level experience, Saint Lucia's Jade Mountain's luxurious sustainable lionfish gourmet dinner is a true hunt-to-table parallel to the premium farm-to-table menus. The resort's chefs prepare a multi-course lionfish tasting each Friday night right on the beach.

Don't miss the ultimate annual Saint Lucia Dive Fest, when Anse Chastanet and Scuba St Lucia celebrate the underwater world with a week of scheduled boat and shore dives, courses, and photographic competitions—and a lionfish eradication day with a chef's culinary demonstration and special dinner feast.

Sustainability Organizations

The Monterey Bay Aquarium aids consumers and chefs by recommending more sustainable seafood choices. I work with Seafood Watch along with several other talented chefs on a Blue Ribbon Task Force to share ideas about seafood sustainability and work toward ensuring a future with a healthy ocean.

The Reef Environmental Education Foundation links the diving community with scientists, resource managers, and conservationists through marine life data collection and related activities. They have a lionfish reporting app to collect current data on sightings and work to promote the REEF Invasive Lionfish derbies in Florida and elsewhere.

Complete eradication of lionfish is unlikely, but there is hope that developing better methods for local removal may be the key to controlling them and mitigating further damage. Research projects are attempting to identify the most efficient and cost-effective methods to fish down lionfish numbers so native fish populations can recover and stabilize.

Lionfish traps have been mostly ineffective in capturing the predators. However, organizations such as the Lionfish University are working with NOAA and other partners to develop new lionfish trapping devices that only attract lionfish so as not to harm native species. These model traps are being tested at much deeper water depths than those divers can access safely. Currently, fish traps are generally prohibited in US federal waters in the Gulf of Mexico and South Atlantic. With a clear preservation mission, this nonprofit is designing prototypical traps that will not damage the fragile ecosystem surrounding the reefs.

Safe Handling

Use care when handling lionfish, as they have up to eighteen venomous spines that can cause painful stings on their dorsal, pelvic, and anal fins. Stings can result in swelling, blistering, dizziness, necrosis, and even temporary paralysis. If stung, immerse the wound in hot (not scalding) water for thirty to ninety minutes and seek medical attention if necessary.

Filleting a lionfish is like filleting any other type of fish except for the need to use caution to avoid the spines located along the dorsal, pelvic, and anal fins. If you put the fish on its side, you can easily hold it by the bony gill plates or soft pectoral fins without getting stuck with a venomous spine. One safety precaution is to wear puncture-resistant gloves. Some also choose to cut off the spines prior to filleting. Use care when doing this as the venomous glandular tissues located within the grooves of the spines are present even at the base of the spine. Furthermore, the venom can remain active in the spines even after the lionfish is dead and stored on ice.

Once you've gotten the spines under control, fillet as you would any other fish, making incisions just behind the spines on the head down to the belly, down the back of the fish near the dorsal spines, and along the bottom of the fish, joining the three cuts together. The skin can be peeled off from the cut closest to the head, or you can continue to cut the fillet away from the body and then cut the fillet from the skin after it has been removed from the body.

Lionfish by the Numbers

1. EAT THEM TO BEAT THEM.
2. Lionfish reach adult size in approximately two years.
3. The largest record lionfish measured a little over 47.7 cm or 19.5 inches and was speared near Islamorada, Florida.
4. In lab studies, lionfish die when water temperatures reach 50 degrees Fahrenheit or 10 degrees Celsius.
5. Lionfish have been visually sighted down to depths of 1,000 feet or 305 meters.
6. Female lionfish are sexually mature and will release eggs when they reach seven to eight inches in length or at approximately one year old.
7. A female lionfish can release between 12,000 and 15,000 unfertilized eggs every four days year round, or approximately two *million* eggs per year, in warm Caribbean waters.
8. A lionfish's stomach can expand to up to thirty times its normal volume.
9. A lionfish can eat prey that is just over half its own body size if it can get its mouth around the prey.
10. Lionfish are known to eat just about every marine creature in their range, including over seventy different fish, invertebrates, and mollusks.
11. A single lionfish may reduce the number of juvenile native fish on any given reef by approximately 79 percent in just five weeks.
12. Lionfish have eighteen venomous spines that are capable of easily penetrating human skin and delivering a very painful sting. Thirteen of these spines are located along the spine in the dorsal fins; there is one short spine in the leading edge of each of the pelvic fins, and there are three short spines in the leading edge of the anal fin. The venom is a protein-based combination of a neuromuscular toxin and a neurotransmitter called acetylcholine. The venom can be denatured (or rendered inert) by applying heat or freezing.
13. Recommended first aid for lionfish stings and envenomation include surfacing safely from a dive, removing any broken spines, and disinfecting the wound, then applying non-scalding hot water for thirty to ninety minutes. Monitor for signs of allergic reaction or shock and react accordingly. Seek medical treatment immediately.
14. There have been *zero* known human fatalities due to a lionfish envenomation.
15. **Yes, you can eat lionfish. They are not poisonous!**

CHEF'S COLLABORATION

Chef...is it safe and is it delicious? This was the thought running through the minds of a dozen of my chef friends when I asked them to collaborate on this sustainable seafood cookbook focused on lionfish by sharing a soulful Caribbean-inspired lionfish recipe. I am humbled by the amazing and rapid responses I received from these chefs, who care greatly about fish cookery and seasonal freshness. These chefs know their fish and are happy to help you eat sustainably.

Sustainable seafood, in simple terms, suggests the intent to make sure that there are enough fish in the sea to maintain a natural balance and sustain life for centuries to come. This includes rivers, streams, lakes, bays, reefs, gulfs, seas, and oceans. It is our hope that our grandkids and their grandkids can enjoy the wonders of nature and have the choice of eating wild fresh fish. You can make that choice today.

I have added recipes to each chapter from a group of very talented chefs and personal friends who are taking a leadership role and making a sustainable statement through their cooking.

Fish cookery is not just for chefs. Anyone can cook fish. It may take some nerve and some practice, but you will learn that it is worth every delicious bite. In fact, some of these chefs had never even tasted lionfish until I introduced it to them. I would like to note here that most of the recipes are for lionfish; nonetheless, these flavors and cooking techniques can be applied to most fish and shellfish. Go for what is freshest.

I introduced lionfish to the culinary world a few years ago in Florida. I had occasionally cooked the fish at Chef Allen's, my restaurant in Miami. But the problem posed by lionfish grew to be even more serious as this invasive species, which was already physically hurting the Florida reefs and coastline, spread to the Bahamas and then throughout the entire Caribbean.

At a recent collaborative benefit dinner for the James Beard Foundation, I chose to serve lionfish to the waiting foodies. They all loved its taste and briny fresh aroma. One guest asked if I had special training in how to carefully prepare this fish, as in Japan you must be certified and strictly licensed to serve the poisonous Fugu blowfish—a risky Japanese delicacy. I explained this is not that kind of fish.

The lionfish is not poisonous, but it does have venom in the dorsal fins, and, when handling the fish, care must be taken to prevent being poked. Luckily, though, these are snipped away by the fishmonger before it comes into the kitchen. In the end, the sense of adventure and the intrigue of this exotic fish make for stimulating conversation and yummy eating.

SPRING

"You cannot get an influence from the cuisine of a country if you don't understand it. You've got to study it."

—Chef Ferran Adria

Dry Season

The Caribbean spring is liberating. The planting season there is at the end of the dry season, leading to young and tender blooming flavors. It is a delicious time of year to try something new. Spring is fresh and green and makes for a green world for soul-searching. I have chosen to begin the adventure of eating lionfish here. The preparations are simple, and the time needed in the kitchen is minimal. It is my intent to get you excited about eating lionfish.

The tropical climate in the Caribbean is divided into two seasons instead of the traditional four. Locals refer to them as the dry season and the wet season. The dry season runs from about February to June. This would compare seasonally to springtime weather on the mainland with its early stages beginning in winter. This is the most ideal time on the islands. Overall, temperatures are in the mid-eighties throughout most of the region, and prevailing northeast trade winds make for comfortable days and cooler evening breezes.

Caribbeans take pride in cooking with local seasonal ingredients, often sourced from their own backyard farm. They live close to the earth in a joyful and simple way. For anyone who adores food and cooking, spring offers refreshing colors and the start of fresh local produce from the garden, farmers' market, and grocery store.

The key is to keep things uncomplicated and let the quality of seasonal ingredients shine through. In the dry season, look for fresh callaloo, kale, Malabar spinach, coconuts, cucumbers, seasoning peppers, green bananas, passion fruit, and fresh young aromatic greens and herbs. Take advantage of spring's abundance.

Spring

SERVES 4

LIONFISH CEVICHE WITH MANGO AND LIME

In honor of springtime, I am offering you a citrusy, light seafood cocktail with a twist. Ceviche is made with raw fish cured in freshly squeezed citrus juices and spiced with chili peppers. Sparkling fresh lionfish makes the best ceviche. Try to keep the lionfish flesh ice cold from the time you purchase the fish until the moment you delight in its taste.

- 1 pound lionfish
- 1 Tablespoon kosher salt
- ½ cup freshly squeezed lemon juice
- ¼ cup freshly squeezed lime juice
- 1 cup white vinegar
- 2 medium jalapeños, minced
- 1 large sweet red pepper, finely julienned
- 1 large firm mango, peeled and finely julienned
- 1 medium red onion, shaved
- 1 teaspoon minced garlic
- ½ Tablespoon freshly ground black pepper
- ¼ teaspoon crushed red chili flakes
- ¼ cup extra-virgin olive oil
- ½ cup freshly picked cilantro leaves

TO PREPARE THE LIONFISH:

Cut the lionfish into short, thin julienned pieces. Place them in a stainless steel bowl and season with 2 teaspoons salt. Pour the lemon and lime juices over the fish, cover, and refrigerate for a half hour. Drain the citrus juices using a colander. Pour the vinegar over the fish while tossing lightly, washing the fish with the vinegar.

TO PREPARE THE CEVICHE:

Using a large bowl, add the jalapeños, peppers, mango, onion, garlic, salt, pepper and chili flakes. Toss lightly and cover with the extra-virgin olive oil. Cover, refrigerate, and let set for at least a half hour. Add the cilantro and mix well before serving.

In spring in the Caribbean, there are many young firm mangos on the trees. When they are fully grown and ripening, the fruit is crunchy and juicy. The mango is slightly sour, like a green apple, and gives the ceviche another layer of favorable interest.

SERVES 4

LIONFISH SOUSCAILLE

This Saint Lucian inspired raw pickled mango and lionfish salad is perfect for summer, and the ginger lime dressing really enhances the flavor of this treat.

- 1 large green mango
- 1 cup cold water
- 2 cloves garlic, crushed
- 1 teaspoon salt
- ½ teaspoon freshly ground ginger
- 1 small Scotch bonnet pepper, minced
- 4 medium West Indian limes, squeezed
- 1 cup cooked lionfish, flaked
- 1 cup organic baby arugula
- 4 small sprigs rosemary
- 1 small carambola, cut through in star-shaped slices

TO PREPARE THE MANGO:

Peel the mango and cut into small cubes. Place them in an earthenware dish. Mix the water, garlic, salt, ginger, Scotch bonnets, and limes together in a small bowl and pour them over the mango. Marinate for at least 45 minutes.

TO FINISH THE LIONFISH SOUSCAILLE:

Carefully add the flaked lionfish to the souscaille mixture and let marinate for 15 minutes.

TO SERVE:

Fill four tall martini glasses with arugula, spoon the souscaille into the glass, and garnish with a sprig of rosemary and a carambola slice.

The Creole translation of Souscaille means "under the house," drunken, or marinated. Feel free to substitute another full-flavored fish, such as mackerel, rainbow runner, or yellowtail snapper.

SERVES 4

GREEN FIG AND SALTFISH (FIG VÉT É LANMOWI)

There is a rich French-based Creole culture on Saint Lucia, and Jounen Kweyole is the official day to celebrate that heritage. During this annual festival, traditional dishes such as Green Fig and Saltfish—the official national dish of Saint Lucia—can be sampled throughout the island with delicious family variations. We use our own freshly salted lionfish instead of the salted dried cod known for centuries in the Caribbean as saltfish.

- 3 small lionfish fillets, prepared in sea salt
- 1 Tablespoon coconut oil
- 1 small onion, diced
- 1 Tablespoon diced seasoning peppers
- 1 clove garlic, crushed
- 2 spring onions, sliced
- 2 small tomatoes, seeded and diced
- 2 to 3 drops hot pepper sauce
- 2 green bananas, boiled until tender and skins have split, then peeled and diced
- 1 ½ Tablespoons mayonnaise
- 1 Tablespoon torn cilantro
- 3 to 4 sprigs fresh celery leaf

TO PREPARE THE LIONFISH AS SALTFISH:

Liberally season the fish fillets with plenty of sea salt to cover completely and allow to cure a minimum of 1 hour and up to 24 hours. Rinse the salt off under running cold water before cooking, then dry the flesh well.

TO COOK THE SALTFISH:

Warm the coconut oil in a heavy pan over medium-high heat. Cook the fish on each side for about 2 to 3 minutes. Remove the fillets and set aside to cool.

TO PREPARE THE GREEN FIG AND SALTFISH:

To the same pan, add the onion, seasoning peppers, garlic, and spring onion, cooking slowly for about 3 to 4 minutes until aromatic. Flake in the cooked fish in thumb-sized pieces and continue to stir in the tomatoes, hot sauce, and diced banana. Remove from heat and fold in the mayonnaise and cilantro.

TO SERVE:

Spoon the mixture onto a large colorful platter and garnish with sprigs of celery leaves.

Green Fig and Saltfish is normally served as a breakfast. To accompany this savory, spicy, and starchy start to the day, I like to top it with a farm-fresh crispy fried egg, sunny-side up. Serve with freshly squeezed carambola juice.

SERVES 4

WEST INDIAN COCONUT SPICED LIONFISH

Coconut oil is one of the few foods that can be classified as a superfood. Its unique combination of fatty acids can have positive effects on your health. Cooking with coconut oil has deep roots in Caribbean culture.

- 2 Tablespoons unsweetened shredded coconut
- 1 Tablespoon ground cinnamon
- ⅛ teaspoon ground cloves
- 1 Tablespoon ground coriander
- 1 teaspoon kosher salt
- 1 teaspoon freshly ground black pepper
- 4 large lionfish fillets
- 4 Tablespoons coconut oil
- 2 Tablespoons minced green onions
- ½ teaspoon fresh grated ginger
- 1 cup freshly made organic carrot juice
- ½ cup freshly squeezed local orange juice
- 2 Tablespoons cold sweet butter

TO PREPARE THE SPICES:
In a small bowl, combine the coconut, cinnamon, cloves, coriander, ½ teaspoon salt, and ½ teaspoon pepper, then blend well.

TO PREPARE THE LIONFISH:
Sprinkle the lionfish liberally with the coconut spice and drizzle with 2 Tablespoons of warmed coconut oil. Cover and refrigerate for 4 hours.

TO PREPARE THE CARROT GINGER SAUCE:
In a medium saucepan over moderate heat, warm the remaining oil, adding the onion and ginger. Cook for 3 to 4 minutes until softened. Mix in the carrot and orange juices and bring the mixture to a simmer. Reduce the liquid to half the volume and season with salt and pepper. Cut the cold butter into 3 pieces and whisk into the sauce.

TO PAN GRILL THE LIONFISH:
Preheat a large seasoned grill pan over high heat. Sear the fish for about 2 to 3 minutes on each side until just cooked through. Place the cooked lionfish on a platter and serve with the sauce on the side.

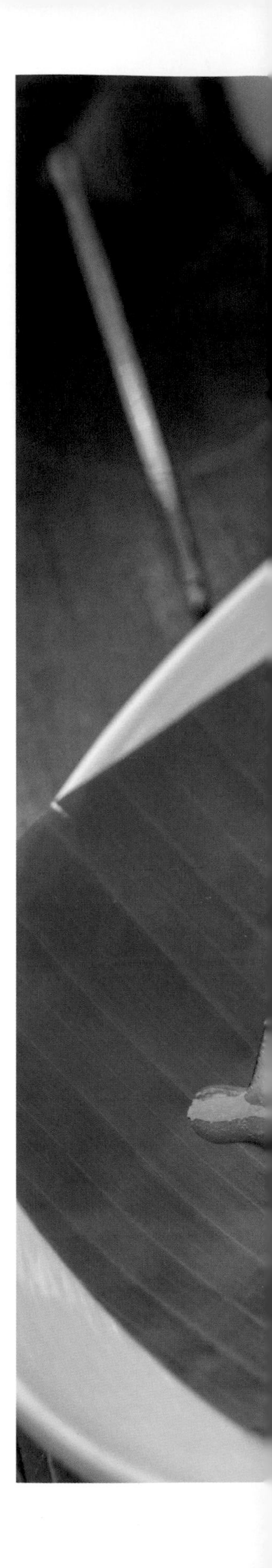

SERVES 4

THAI BASIL BAKED LIONFISH

The Thai basil has the distinctive trait of an anise or licorice-like flavor and is slightly spicy. It has narrow shiny green leaves with purple flowers. Here it is used minced in a marinade and garnish for the finish. This would pair well with steamed jasmine rice.

- 4 large lionfish fillets
- 2 large lemons, zested and juiced
- 4 sprigs fresh Thai basil, 2 sprigs minced
- ½ teaspoon mint leaves
- ½ teaspoon freshly ground white pepper
- 1 teaspoon sea salt
- 1 teaspoon minced garlic
- 2 Tablespoons olive oil

PREPARING THE FISH:
Wash the fish in cold water and drain. In a stainless steel bowl, combine the lemon juice, zest of the lemons, Thai basil, mint, white pepper, salt, garlic, and olive oil. Place the fish into the marinade.

BAKING THE LIONFISH:
Preheat the oven to 400 degrees. Place the fish in an ovenproof dish with marinade and bake, moistening the lionfish with marinade occasionally. Bake for approximately 5 to 6 minutes until flesh begins to flake.

TO SERVE:
Remove from oven and garnish with fresh Thai basil sprigs.

SERVES 4

LIONFISH WITH MALABAR SPINACH PESTO

We grow our own Malabar spinach on our Emerald Estate organic farm in Saint Lucia. Malabar spinach has very fleshy, thick leaves that are juicy and crisp with tastes of citrus and pepper. Malabar spinach is high in Vitamin A, Vitamin C, iron, and calcium; it has a high amount of protein for a plant and is used widely in the West Indies.

- 4 large lionfish fillets
- 1 teaspoon lemon zest
- 1 teaspoon sea salt
- ½ teaspoon crushed red pepper flakes
- 4 cloves garlic
- ¼ cup toasted pine nuts
- ½ cup fresh basil leaves
- ¼ cup fresh parsley leaves
- 1 cup Malabar spinach leaves
- ½ cup grated parmesan
- 1 teaspoon lemon juice
- 1 ¼ cup olive oil

TO PREPARE THE LIONFISH:

Season the lionfish with lemon zest, a little sea salt, and pepper flakes. Drizzle with olive oil. In a heavy bottomed pan, over medium-high heat, sauté the lionfish until nicely browned on both sides.

TO PREPARE THE PESTO:

In a food processor, combine the garlic, nuts, basil, parsley, spinach, parmesan, and lemon juice. Pulse until ground into a fine paste. With the machine running, drizzle in the olive oil until pesto reaches desired consistency. Season with salt and pepper flakes. Place in a mason jar and store until needed.

TO SERVE:

Spoon a generous spoonful of pesto onto each plate. Arrange the lionfish in the center.

Callaloo, which is comprised of the green, leafy tops of the dasheen plant, is interchangeable with Malabar spinach. At home, if needed, substitute leafy green spinach. This pesto is brightly flavored with fresh herbs and the addition of lemon juice.

SERVES 4

GREEN SPICED LIONFISH WITH PLANTAINS

Plantains are frequently eaten here in Saint Lucia and throughout the Caribbean. Cooked green plantains are packed with minerals, which are very important for the body. The fruit contains lots of potassium, which helps with stress and boosts the metabolism.

- 1 teaspoon fennel seed
- 1 teaspoon cumin seed
- 3 Tablespoons dried lemon zest
- 2 Tablespoons smoked paprika
- 1 ½ Tablespoons granulated garlic
- 1 Tablespoon dried thyme
- 2 teaspoons kosher salt
- 2 teaspoons cane sugar
- ¼ cup freshly chopped parsley
- 3 small green plantains
- 2 Tablespoons coconut oil
- 4 large lionfish fillets
- 1 large lemon, sliced in thin wheels

TO PREPARE THE GREEN SPICE:

In a small dry pan over medium heat, toast the fennel seeds and cumin seeds until aromatic, for about 2 minutes. Cool and crush in a mortar and pestle. Add in and work together the lemon zest, paprika, garlic, thyme, salt, sugar, and parsley. This will make 1 cup of spice.

TO PREPARE THE PLANTAIN:

Peel the plantain and slice crosswise into small coins. In a small pot, boil the plantain in 1 quart of water with 2 Tablespoons of the green spice. Cook until softened, about 6 to 7 minutes.

TO PREPARE THE LIONFISH:

Season the lionfish with the green spice and let set covered for 30 minutes. Warm the coconut oil in a heavy-bottomed pan. Carefully slip the lionfish in and pan fry until golden brown, about 2 minutes on each side.

TO SERVE:

Place cooked fish on platter with green plantains and garnish with lemon wheels.

SERVES 4

PAN ROASTED LIONFISH WITH PASSION FRUIT BUTTER

The sustainably caught lionfish is heart-healthy and high in omega-3 fats. When cooked to a white, translucent stage, its firm, light-textured taste is like a small grouper fillet or mahi-mahi, but with one notable difference: a hint of butter.

- 2 large shallots, minced
- 1 cup white wine
- 2 Tablespoons rice wine vinegar
- 1 large passion fruit, fresh pulp
- ¼ pound butter, cut into small cubes—keep cold
- 1 teaspoon kosher salt
- 4 large lionfish fillets
- 1 teaspoon freshly chopped tarragon
- 3 black peppercorns
- 1 Tablespoon butter, softened
- 1 cup organic arugula

TO PREPARE THE BUTTER SAUCE:

In a saucepan, sweat shallots in butter until translucent, about 5 minutes. Add wine and vinegar and bring to a boil. Lower to simmer and reduce for 15 minutes until only 2 Tablespoons of liquid remain.

TO MAKE THE PASSION FRUIT BUTTER SAUCE:

Stir in the passion fruit and continue to reduce until syrupy for about a minute. With the pan on very low heat, whisk in the cold butter pieces—add one piece at a time, waiting until each piece melts into the sauce. Season with salt and strain through a fine sieve. Keep warm until ready to serve.

TO MAKE THE LIONFISH:

Season the fish with tarragon, salt, and pepper. In a heavy pan over medium heat, melt the remaining butter. Cook the lionfish on each side for 2 to 3 minutes.

Place organic arugula on the plate, arrange the lionfish, and serve with passion fruit butter on the side.

SERVES 4

SPICY LIONFISH TACOS WITH GREEN MANGO CHOW CHOW

Fish tacos are served throughout the Caribbean, and Lionfish is set to be the taco fish of the moment. The chow chow calls for mature, ripe, green mangos, which add crunch and character. This sexy lionfish taco will be a big hit at your next beach party.

- 1 large green mango, diced
- 1 large seasoning pepper, diced
- 1 small red onion, diced
- 3 Tablespoons cilantro, chopped
- 1 Tablespoon lime juice
- 1 teaspoon ground cumin
- ½ teaspoon chili powder
- ½ teaspoon garlic powder
- ½ teaspoon allspice
- ½ teaspoon crushed red pepper flakes
- 1 teaspoon sea salt
- 8 small lionfish fillets
- 3 Tablespoons coconut oil
- 8 small white corn tortillas
- ½ cup red cabbage, shredded
- 1 large avocado, sliced in 8 wedges
- 2 Tablespoons sour cream
- 1 large lime, cut in thin wedges

TO PREPARE THE MANGO CHOW CHOW:

In a medium stainless steel bowl, combine the mango, seasoning pepper, red onion, 2 Tablespoons cilantro, lime juice, and 1 teaspoon of spice mix you'll make in the next step.

TO PREPARE THE FISH:

In a small bowl, combine the cumin, chili, garlic, allspice, salt, and red pepper. Use spices to season the lionfish. Heat the coconut oil in a heavy pan and fry the lionfish in batches until golden brown.

TO SERVE:

In a hot cast-iron pan, toast the tortilla shells. Fill each with the lionfish, cabbage, lettuce, and mango chow chow. Top with a dollop of sour cream, then dust with remaining spice mixture and cilantro. Serve with lime wedges.

SERVES 4

LIONFISH TARTINE

There are dozens of varieties of Caribbean avocado. The outer shell can range from smooth light green to dark green alligator-like skin. They are known locally as alligator pears. The interior flesh is aromatic and yellow-green in hue. Be sure to use a ripe avocado that is soft and giving to a light squeeze. This tartine is a classy yet simple open-face sandwich on freshly toasted country style bread.

- 4 large lionfish fillets
- 1 Tablespoon chopped cilantro
- 1 teaspoon sea salt
- ¼ teaspoon crushed red pepper flakes
- 2 Tablespoons butter
- 1 Tablespoon capers with juice
- 2 teaspoons chopped parsley
- 4 thick country bread slices
- 1 small avocado, scooped
- 1 large heirloom tomato
- 1 small lime, cut in wedges

TO PAN FRY THE LIONFISH:

Season the fish with cilantro, salt, and red pepper. In a heavy pan over medium heat, melt the butter. When the butter is just about completely melted, add the lionfish, cooking the first side for 2 to 3 minutes. Turn the fish over, and, with a spoon, glaze fish with the cooking liquid. Add the capers and parsley, continuing to baste the fish until cooked through.

TO PREPARE THE TARTINE:

Toast the bread slices to desired doneness. Divide the avocado between the toasts and lightly smash with the prongs of a fork. Arrange the tomato and lionfish on each tartine. Top with caper pan juices and garnish with lime wedges.

SERVES 4

LIONFISH SUQUET
CHEF'S COLLABORATION: José Andrés

Serving lionfish at our restaurants is a win-win-win situation... first, they are so delicious. They have light flaky flesh and can be prepared many ways. The next win is that by making dishes with lionfish, we are helping the ecosystem of the Bahamas' waters. Lionfish are invasive and incredibly destructive, and they don't have any natural predators in the Caribbean. They are predatory toward other fish and shellfish, especially some that are very important to local fishermen. So that's the third win—by serving lionfish, we are supporting fishermen and the local community. We will buy lionfish at a good price for the fishermen; it is a difficult fish to catch, since it requires hunting with a spear, which is another environmental benefit, since catching them doesn't damage the reef. Overall, this is one of the best dishes that we serve anywhere—it supports the local economy, helps the environment, and tastes incredible!

Twice named to *Time* magazine's "100 Most Influential People" list and awarded "Outstanding Chef" and "Humanitarian of the Year" by the James Beard Foundation, José Andrés is an internationally recognized culinary innovator, *New York Times* bestselling author, educator, television personality, humanitarian, chef/owner of ThinkFoodGroup, and the founder of World Central Kitchen.

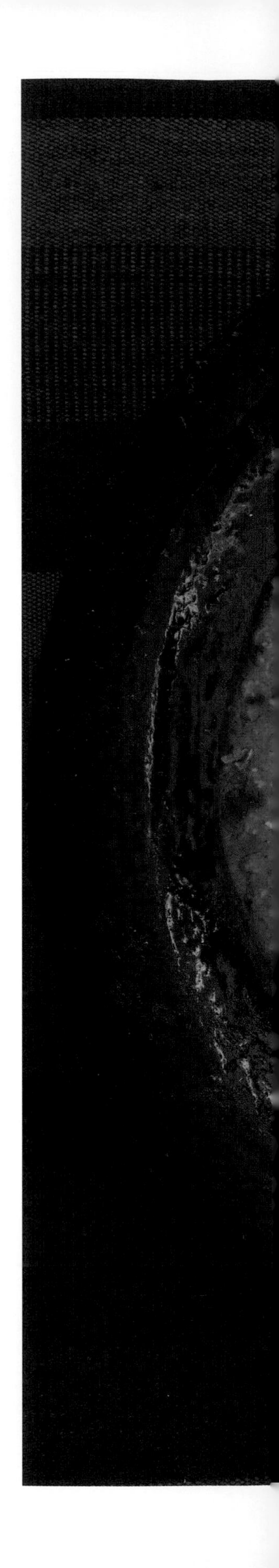

- ¼ cup extra-virgin olive oil
- 1 large onion, finely diced
- 1 15.5-ounce can crushed tomatoes
- 2 teaspoons smoked pimentón
- 1 bay leaf
- 4 slices of old bread
- ½ cup almonds
- 2 garlic cloves
- ¼ cup parsley, destemmed
- 2 to 3 saffron pistils
- 2 large white potatoes, peeled and cut into half-inch cubes
- 4 cups fish stock
- 1 pound lionfish fillets, cut in 2-inch cubes
- 8 large shrimp, cleaned
- Minced parsley for garnish

START BY MAKING A SOFRITO:

Heat the oil in a medium saucepan over medium-low heat. Add the onions and cook, stirring occasionally, until they are soft and golden brown, about 4 to 5 minutes. You want the onions to caramelize; if they start to get too dark, add ½ teaspoon of water to keep them from burning. Stir in the crushed tomatoes, pimentón, and bay leaf and cook for another 20 minutes over medium heat. You'll know the sofrito is ready when the tomatoes have broken down and deepened to an almost rusty color and the oil has separated from the sauce. Discard the bay leaf.

TO MAKE THE PICADA:

Toast the bread and sauté the almonds in a sauté pan until they are aromatic. In a food processor, pulse together the garlic, parsley, and saffron or alternatively, crush in a mortar and pestle, starting with the saffron to release its aroma. Add in the bread and almonds, pulsing to a smooth paste. Set aside.

TO PREPARE THE SUQUET:

Add cubed potatoes and fish stock to the sofrito and cook for 10 to 15 minutes or until the potatoes are almost tender. Taste the stock and add salt as needed—you'll need more or less depending on how salty your fish stock is. Add the lionfish and cover the pot; cook for 2 to 3 minutes. Add the shrimp and cook for another 2 to 3 minutes. Stir the picada into the stew to thicken it and cook for a few more minutes, stirring.

TO SERVE:

Taste and adjust for salt, then serve this juicy fish stew hot with a sprinkle of parsley on top.

SERVES 4

BREADFRUIT GNOCCHI WITH LIONFISH SOFRITTO

CHEF'S COLLABORATION: Nina Compton

This is a modern taste of the Caribbean Boil and Fry style of comfort food, which is a popular way to cook ground provisions. The breadfruit is transformed into airy pillows of gnocchi and fried with onions, tomatoes, saltfish, and herbs. To prepare the lionfish for this recipe, as saltfish normally would be used, use a cup of coarse kosher salt to cover and cure one pound of lionfish overnight in a refrigerator. The next day, rinse the salted lionfish well.

BREADFRUIT GNOCCHI

- Yields 120 (1-inch) gnocchi; recipe serves 10 to 12
- 1 medium (5-pound) breadfruit, or substitute Yukon gold potatoes
- 1 Tablespoon salt
- 5 large egg yolks
- ½ pound all-purpose flour

SOFRITTO

- ½ cup salt, for the cooking water
- 4 Tablespoons coconut oil, or may substitute olive oil
- 3 medium Roma tomatoes, diced medium
- ½ medium Spanish onion, diced
- 1 poblano pepper, diced small
- 1 bunch scallions, finely sliced, reserving 1 Tablespoon for garnish
- 1 pound of salt cured lionfish, broken into large pieces
- ¼ cup rum
- 1 Tablespoon chopped parsley

TO MAKE THE BREADFRUIT GNOCCHI: place the breadfruit (or the potatoes) in a large stockpot with enough cold water to cover. Add the salt and cook over medium heat for about 20 to 30 minutes, until tender. Drain the water and let the breadfruit cool enough to handle, then peel and core. Transfer to a large bowl and mash the breadfruit, using a potato masher or a fork, until it is smooth and there are no lumps remaining.

TO MAKE THE GNOCCHI DOUGH: Stir in the egg yolks, then slowly add the flour, mixing gently. Do not overwork the dough. Once the mixture comes together and is not too soft or too tight, it is ready to roll out.

TO SHAPE THE GNOCCHI: On a floured surface, divide the dough into 6 parts, and roll each part into half-inch diameter logs, about 20 inches long. Cut the logs into one-inch pieces. Transfer the gnocchi to two or more parchment-lined sheet pans and arrange them in one layer, not touching. Refrigerate the gnocchi at least one hour before cooking. If you do not wish to cook the gnocchi immediately, or you do not wish to use all the gnocchi at this time, the remainder can be sealed in a zip-top plastic bag and frozen for up to one week before cooking.

TO COOK THE GNOCCHI:

Fill a separate large stockpot half full of water, add ½ cup salt, and check the water—it should taste as salty as the ocean. Bring to a boil over high heat. Add the gnocchi and cook until they float, about 1 minute, then remove using a slotted spoon.

TO PREPARE THE LIONFISH SOFRITTO:

In a medium sauté pan over medium heat, add the coconut oil, tomatoes, onion, peppers, scallions, and the lionfish; cook for 3 minutes, stirring. Remove the pan from the heat and set aside.

TO FINISH AND SERVE:

Add the cooked gnocchi and a splash of the cooking water (since the starch from the gnocchi water acts a binder) to the sauté pan with the tomato soffrito. Remove the pan from the heat and add the rum, then return the pan to medium heat and cook 5 minutes, stirring occasionally, until the liquid has reduced almost completely. Serve with a sprinkling of chopped parsley and finely sliced scallions.

Saint Lucia Culinary Ambassador and native, Nina Compton is chef/owner of Compère Lapin in New Orleans, which has received a great deal of critical acclaim, including a rave review in *The New York Times*, Top 10 Winner of *Playboy*'s Best New Bars in America, and "Best New Restaurant" nods from *New Orleans Magazine* and the *Times-Picayune*. The talented chef was awarded the "Best Chef South" by the James Beard Foundation and has been named one of *Food & Wine* magazine's "Best New Chefs."

Chef Bradley Kilgore opened Alter in the art-focused Wynwood district of Miami. There he offers Progressive American cuisine, highlighting indigenous Floridian ingredients. Alter was named a semi-finalist for the prestigious James Beard Awards in the category of Best New Restaurant, and Chef Kilgore has twice been named a semi-finalist Rising Star chef. The James Beard Foundation named him a finalist as Best Chef South. Chef Kilgore was named Best New Chef in America by *Food & Wine* magazine.

He is currently the culinary director for the Adrienne Arsht Center and has a signature restaurant, Brava by Brad Kilgore. He recently opened two new restaurants in the Design District, Kaido, an Asian and Japanese Cocktail Lounge and Restaurant, and Ember, a Wood Fire New American Grill.

SERVES 4

LIONFISH SASHIMI WITH SHIRO PONZU, AVOCADO MOUSSE, AND YUZU KOSHO

CHEF'S COLLABORATION: Brad Kilgore

I don't have the time or skills to go and catch the invasive fish myself, but I can serve it at my restaurant. Guests can do their part by ordering it, too. Luckily, it's quite delicious, but hopefully soon it will be eradicated!

PONZU

- 1½ Tablespoons lemongrass, julienned
- 1 teaspoon ginger, julienned
- 2 leaves kaffir lime leaves, julienned
- 20 each coriander seeds
- ½ teaspoon serrano chili
- 1o fresh basil leaves
- 2 lemons, zested
- 3 Tablespoons lemon juice
- 3 Tablespoons yuzu
- 1 cup white soy

AVOCADO

- 1 avocado, peeled and pitted
- 2 Tablespoons lime juice
- ½ teaspoon sea salt
- 2 Tablespoons olive oil

SASHIMI

- 4 large lionfish fillets
- 2 teaspoons yuzu kosho
- ½ teaspoon coarse sea salt
- 5 to 6 fresh edible flowers
- 2 Tablespoons high quality, fruity olive oil

TO PREPARE THE SHIRO PONZU:

Using a mortar and pestle, start to grind the lemongrass, ginger, kaffir lime leaves, coriander seeds, serrano, basil, and lemon zest. Continue to crush the mixture, moistening with lemon juice and yuzu. Stir in the soy and let marinate without straining for 30 minutes, then strain through a fine mesh and set aside.

TO PREPARE THE AVOCADO MOUSSE:

Place the cleaned avocado, salt, and lime in a Vitamix blender. Pulse together to blend. With the machine running, drizzle in the olive oil. Spoon into a piping bag and chill for 30 minutes.

TO FINISH AND SERVE:

Slice the lionfish sashimi style, arranged decoratively and concisely in a semicircle on a large glass platter. With a clean modern hand, dot and dollop the avocado mousse and yuzu kosho around the lionfish. Sprinkle with the salt and glaze with ponzu. Garnish with petals of flowers, finishing with a drizzle of your favorite fruity olive oil.

SUMMER

"This is my advice to people: learn how to cook, try new recipes, learn from your mistakes, be fearless, and above all, have fun."

—Julia Child

Wet Season

Summertime in the Caribbean is when the tropical isles are bursting with flavorsome lush fruits. When you eat fruit off the tree, it tastes ten times better than normal. It is warmed by the sun, which gently cooks the fruit and allows its natural sweetness to shine. Summer brings mango, pineapple, soursop, guava, papaya, and avocado. In fact, the Caribbean is home to some of the most delicious mangos in the world. Saint Lucia is dripping with bright, colorful mangos of vastly different flavors and textures. Celebrate with us the sweet joys of exotic tropical fruits, fish, and the luscious crunch of vegetables.

The wet season, which runs from July to December, extends well past summer. Part of the wet season is the hurricane season that starts in June and ends in November. Thankfully, most storms pass to the north of Saint Lucia when they are rolling across the Caribbean Sea. The frequency and intensity of Caribbean storms vary greatly from year to year. Even during the rainy period, however, the precipitation range fluctuates significantly. The windward sides of islands with similar mountainous landscapes to Saint Lucia receive a lot of rain, whereas the leeward sides can have very dry conditions. Flat islands receive slightly less rainfall, but their precipitation pattern tends to be more consistent.

Pairing summertime fruits and vegetables with fish is fun. Both cooks and chefs enjoy the flexibility of summer style with grilling, barbecuing, cooking with cast iron, and making sandwiches and salads. Use these recipes to inspire your cooking, but don't get tied down to them. All the fish recipes in this cookbook are for lionfish for good reason: however, each recipe will work deliciously with other local and seasonal fish such as mahi-mahi, kingfish, wahoo, snapper, or yellowtail.

At times, it seems as if time has stood still in Saint Lucia, harkening back to a slower pace of life. The island is ringed by settlements, many of which had their origins as fishing villages, where food is central to daily life. In the quaint fishing villages of Deanery, Choisuel, Soufriere, Canaries, and Anse le Rey, the culture is focused on the sea. Each morning at the break of dawn, small, narrow, colorful boats skim the coastal waterfronts. Four-man teams work close to the shore throwing hand drawn nets, then hoist the sea's harvest into their boats. Capturing lionfish is a bit more like hand-to-hand combat. Most of the lionfish apprehended are speared by locals snorkeling near the many reefs. The fish is then brought to market daily.

The day repeats itself with the rising sun the following morning. There are ways to fish sustainably, allowing us to enjoy seasonal catches while ensuring that seafood populations remain for the future.

Summer

SERVES 4

MANGO, AVOCADO, AND LIONFISH SALAD

This beautifully composed crunchy salad is drizzled with a hot, sweet, and sour dressing. It makes a refreshing dish for a Sunday brunch.

- 1 cup white wine
- ½ cup diced onion
- ½ cup diced celery
- ½ teaspoon kosher salt
- 4 sprigs cilantro, stems and leaves separated
- 4 large lionfish fillets
- 1 small cucumber, peeled, seeded, and sliced
- 1 cup vine ripe cherry tomatoes, halved
- ½ cup bean sprouts
- ⅓ cup rice vinegar
- 2 Tablespoons freshly squeezed lime juice
- 1 small red Thai chilies, seeded and minced
- 1 teaspoon sugar
- 1 small ripe mango, peeled and sliced
- 1 large avocado, peeled and sliced

TO PREPARE THE LIONFISH:

In a small low-sided pot, cook together the wine, onion, celery, salt, and cilantro stems with 3 cups of water. After it simmers for 10 minutes, add the lionfish to poach for 6 to 7 minutes. Remove from the poaching liquid and cool. Refrigerate until ready to use.

TO PREPARE THE SALAD:

In a medium bowl, toss the cucumber, tomatoes, and bean sprouts together. Cover and refrigerate for at least 1 hour, or up to 4 hours. In a small bowl, combine the vinegar, lime juice, chilies, and sugar. Stir until the sugar dissolves.

TO SERVE:

Arrange the salad vegetable mixture, avocado, and mango on four salad plates. Carefully flake the lionfish over the salads. Drizzle with the dressing and garnish with the cilantro leaves.

SERVES 4

LIONFISH COCONUT MILK CEVICHE

Coconut milk is derived from the flesh of the coconut. It is not the liquid that can be drained out from a coconut that has been punctured, although many people assume this. Getting coconut milk from a coconut requires some processing. When a coconut is cut open, the flesh can be found all along its inner walls. A ripe coconut, ideal for processing into coconut milk, should have thick, creamy white flesh. Coconut milk is used as a cooking base in many recipes of the West Indies.

- 1 pound lionfish, medium diced
- ¼ cup West Indian lime juice
- 1 Tablespoon minced ginger
- 1 teaspoon minced garlic
- 2 Tablespoons minced green onion
- 1 Tablespoon sesame oil
- 2 Tablespoons minced cilantro
- 1 Tablespoon soy sauce
- 1 cup coconut milk
- 3 Tablespoons fresh coconut, toasted

TO PREPARE THE CEVICHE:
In a large, stainless steel bowl, combine the lionfish with the lime juice. Separately combine the ginger, green onion, and garlic. In a small pan, heat the sesame oil until just ready to smoke and drizzle into the herb mixture. Add the cilantro, soy sauce, and coconut milk into the fish along with the herb oil. Chill well for 2 hours. When ready to serve, garnish with toasted coconut.

West Indian limes are like key limes. They are small and ripen to a pale-yellow hue. The beauty of these limes is that they are very high in acid (sourness) making them a perfect citrus for ceviche. It releases a pleasant flower blossom aroma when squeezed. I find them handy for making refreshing rum mojito cocktails too.

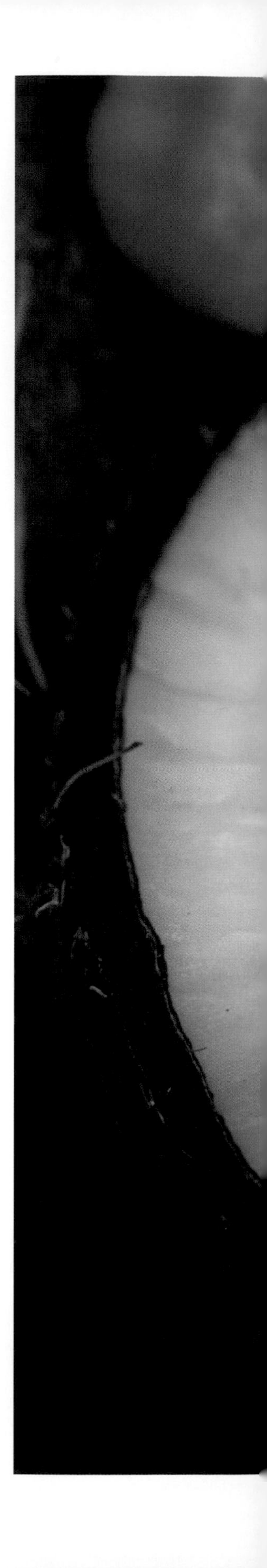

SERVES 4

GRILLED LIONFISH, PAPAYA, PINEAPPLE, AND KALE

Dating back to prehistoric times, the papaya is native to South and Central America. Today, the fruit grows all over Saint Lucia and throughout the Caribbean. This sweet, smooth-fleshed fruit is a great addition to salads, delicious dried, and perfect as a snack. Inside, the flesh is a light orange-pink color and a central cavity filled with small edible black seeds.

- 1 bunch Tuscan kale, stalks removed and discarded, leaves thinly sliced
- 1 large lemon, juiced
- ¼ cup extra-virgin olive oil
- 1 teaspoon kosher salt
- 2 teaspoons honey
- ½ teaspoon freshly ground black pepper
- 4 large lionfish fillets
- 2 sprigs fresh thyme
- 1 small papaya, diced
- ½ small pineapple, diced
- ¼ cup toasted pumpkin seeds
- ¼ cup shaved Parmigiano-Reggiano

TO PREPARE THE KALE:
In large serving bowl, add the kale, half the lemon juice, and a drizzle of oil and season with kosher salt. Massage until the kale starts to soften and wilt, 2 to 3 minutes. Set aside.

TO PREPARE THE DRESSING:
In a small bowl, whisk the remaining lemon juice with the honey and freshly ground black pepper. Stream in 3 Tablespoons of oil while whisking until a dressing forms.

TO PREPARE THE GRILLED LIONFISH:
Preheat the grill. Season the lionfish with thyme, salt, and pepper. Grill the lionfish about 2 to 3 minutes on each side.

TO SERVE:
Arrange the seasoned kale on a platter. Add the papaya, pineapple, grilled lionfish, and pumpkin seeds. Drizzle with dressing and garnish with parmesan.

SERVES 4

CARIBBEAN "FISH AND CHIPS" LIONFISH AND BREADFRUIT

The breadfruit is a common food among the islands due to its versatility and ability to grow in almost any soil condition. The tall evergreen tree produces a hearty bounty two to three times a year. Originally, the breadfruit was brought to the Caribbean from the Pacific as a reliable source of food to feed the slaves. Today, the breadfruit has become a treasured part of Caribbean food and culture and is a favorite staple of many people. The Piton beer is a locally brewed beer Saint Lucian named after the world heritage Pitons—a pair of volcanic spires.

Caribbean cooking uses numerous rooty and starchy vegetables. They make a great variety of sweet, nutty flavored chips. Cassava, yuca, dasheen, malanga, sweet potato, yams, green banana, and plantains can each be prepared as chips as described in the recipe.

- 1 cup Piton beer, or another pilsner beer
- 1 Tablespoon spiced rum
- 1 ¼ cups all-purpose flour
- 2 teaspoons kosher salt
- 2 to 3 cups vegetable oil, for frying
- 1 small breadfruit
- 1 teaspoon curry powder
- 8 small lionfish fillets

TO PREPARE THE BEER BATTER:

Whisk together the beer, rum, 1 cup of flour and 1 teaspoon of salt in a large bowl. Set aside.

PREPARE THE OIL FOR FRYING:

Fill a Dutch oven with vegetable oil, so that oil comes up a little less than halfway up the pot. Place over low heat and heat to 375 degrees.

TO PREPARE THE BREADFRUIT:

Wash and peel the breadfruit. Slice as thinly as possible. Add the breadfruit slices to the hot oil, a few at a time, and cook until golden. Remove from oil and drain excess oil on paper towels. Sprinkle lightly with the remaining salt and curry powder.

TO PREPARE THE FISH:

Place the remaining flour in a resealable bag and set aside with the beer batter. To fry the fish, place fillets in the bag of flour to lightly coat, then dip into beer batter, and add to the oil. Fry until golden brown, turning fish occasionally, about 4 to 5 minutes depending on the size of the fillets. Remove from oil and drain on a paper towel-lined plate for a minute before plating.

TO SERVE:

On a craft paper-lined basket, serve fried fish over a generous portion of breadfruit chips with wedges of lime for garnish and malt vinegar on the side.

SERVES 4

GRILLED LIONFISH AND ROMAINE WITH GINGER VINAIGRETTE

A distant cousin to bamboo and kin to turmeric and cardamom, ginger grows well on our Emerald Estate. Fresh ginger root is a must-have in every Caribbean kitchen. The fresh root is used grated to flavor endless savory dishes.

- 8 small lionfish fillets
- 2 small romaine lettuce hearts
- 2 teaspoons sea salt
- 1 teaspoon freshly ground black pepper
- 1 cup coconut oil
- 2 ounces fresh lime juice
- 2 ounces sherry vinegar
- 2 Tablespoons freshly grated ginger
- 2 Tablespoons boiling water

TO PREPARE THE LIONFISH:

Preheat the grill. Season both the lionfish and the romaine hearts with salt and pepper. Drizzle with a little of the warmed coconut oil. Over medium heat, grill the fish carefully until done on both sides. Continue to grill the romaine hearts until nicely charred.

TO PREPARE THE VINAIGRETTE:

Combine remaining oil, lime juice, vinegar, and ginger in a food processor for about 30 seconds. Then with the motor running add the water and process for another 10 seconds.

TO SERVE:

Arrange the romaine in the center of the platter. Top with grilled lionfish and drizzle with ginger vinaigrette.

SERVES 4

MANGO LIONFISH WITH MINT CHUTNEY

Serve this lionfish with red quinoa pilaf. Add a bit of nutmeg, clove, and ginger to deliver that delicious kick that makes Saint Lucian cuisine so unique in the Caribbean.

- 2 cups mint leaves, packed well
- ¼ cup fresh lime juice
- 2 Tablespoons cilantro leaves
- 3 Tablespoons chopped shallots
- 2 Tablespoons minced green chilies
- 1 Tablespoon brown sugar
- 2 teaspoons salt
- ½ teaspoon freshly ground black pepper
- 4 large lionfish fillets
- 2 Tablespoons olive oil
- 1 ripe mango, peeled and thickly sliced

TO PREPARE THE MINT CHUTNEY:
Place mint, lime juice, cilantro, shallots, green chilies, brown sugar, and 1 teaspoon salt into a blender and pulse twice and then let run for 30 seconds continuously. Pour into a mason jar, cover, and refrigerate for 1 hour before serving.

TO PREPARE THE LIONFISH:
Preheat the grill. Season the lionfish with the remaining salt and the pepper. Brush the olive oil on the fish and on the grill. Grill the lionfish for 2 to 3 minutes on each side until golden brown.

TO PREPARE THE MANGO:
Brush the olive oil on the mango slices and on the grill. Grill the mango for 2 to 3 minutes on each side until golden brown.

TO SERVE:
Arrange the lionfish on a clear platter. Spoon the mint chutney onto the plate. Garnish with hot grilled mango.

SERVES 4

BBQ LIONFISH WITH ORANGE AND ALMOND SLAW

Everyone loves a summer BBQ, and fantastic coleslaw can be a game changer. This tangy coleslaw recipe with crunchy and aromatic additions of fennel, orange, and almonds might even steal the show. Other fish that I like for BBQ include mahi-mahi, king mackerel, and mullet.

- 1 large bulb of fennel, thinly sliced
- 1 small cabbage, shredded
- 1 clove garlic, minced
- 2 large oranges, peeled and sliced
- 1 small red onion, thinly sliced
- ¼ cup Caribbean almonds
- 1 teaspoon kosher salt
- ½ teaspoon freshly ground black peppercorns
- 3 Tablespoons olive oil
- 6 leaves fresh basil, torn
- 3 Tablespoons fresh lemon juice
- ½ teaspoon crushed coriander seed
- 4 large lionfish fillets

TO PREPARE THE ORANGE SLAW:

In a small bowl, combine the fennel and cabbage with garlic, slices from 1 orange, onion, almonds, ½ teaspoon salt, ¼ teaspoon black pepper, 2 Tablespoons olive oil, and freshly torn basil. Cover and refrigerate for a half hour.

TO COOK THE LIONFISH:

Heat charcoal BBQ grill and brush it with a Tablespoon of the oil. Season the lionfish with remaining salt, pepper, and crushed coriander seed. Place the fillets into direct heat and grill the first side for 2 minutes and then carefully turn them over and cook the second side for another 2 to 3 minutes until just cooked through.

TO SERVE:

Spoon 2 to 3 Tablespoons of the orange slaw onto the plates. Place the BBQ lionfish on each mound. Garnish with remaining orange slices.

SERVES 4

STEAMED LIONFISH AND ALMOND ANCIENT GRAINS

All whole grains in the larger sense are "ancient"—they all can trace their roots back to the beginnings of time. Ancient grains are certainly more nutritious than refined grain products; healthy whole grains need not be exotic, as they are already delicious. Feel free to substitute sorghum, freekeh, barley, millet, or amaranth in this recipe to vary the flavor.

- 3 Tablespoons clarified butter
- 3 Tablespoons slivered almonds
- 3-inch piece cinnamon stick
- 3 whole cardamom pods
- 5 whole black peppercorns
- ¼ teaspoon saffron
- 4 large lionfish fillets
- 2 teaspoons kosher salt
- ½ cup red quinoa
- ½ cup farro
- 3 Tablespoons golden raisins
- 3 Tablespoons chopped scallions

TO PREPARE THE NUTS AND SPICES:

In a large saucepan, heat the butter over medium-low heat. Add the almonds and continuously shake the pan until the nuts turn golden brown. With a slotted spoon, transfer the nuts to a small bowl. To the same pan and butter, add all the whole spices and stir constantly for 2 minutes until aromatic.

SEASON THE LIONFISH:

Soak the saffron in 1 Tablespoon of hot water. Brush the lionfish with the saffron water and season with a little salt. Cover and set aside.

TO PREPARE THE ANCIENT GRAINS:

Stir the ancient grains into the spice mixture until well coated. Add remaining salt and the water and bring to a boil. Decrease heat, cover the pan, and simmer for about 10 minutes. When most of the liquid has been absorbed, add the lionfish on top of the rice to steam covered for another 5 minutes.

TO FINISH:

Remove from the heat and set the steamed lionfish on a warm service plate. Fluff the grains with a fork and stir in the prepared nuts, raisins, and scallions. Cover and let set for 5 minutes before serving.

SERVES 4

LUCIAN POT FISH STEW

Pot fish refers to fish pot or fish trap. Pot fishing is a time-honored tradition in Saint Lucia. Often it takes both father and son for a collaborative effort to pull up the traps from more than thirty feet below the surface. The catch is small today in many cases, because the lionfish are eating many small local fish species. Pot fish often include several varieties of small fish cooked together. However, fish stew is probably the most delicious way locals devour them. The stew is often served with rice and ground provisions.

- 4 large lionfish fillets
- 4 small snapper fillets, skin on
- 1 lime, juiced
- ¼ cup coconut oil
- 1 medium red onion, diced
- 1 teaspoon minced garlic
- ½ teaspoon minced ginger
- 1-inch piece cinnamon stick
- 2 sprigs fresh thyme
- 1 bay leaf
- 1 seasoning pepper, sliced
- 3 medium tomatoes, diced
- ½ cup white wine
- 1 sweet bell pepper, sliced
- 3 green onions scallion, chopped
- 12 fresh okra, sliced
- 2 sprigs cilantro, chopped
- 1 teaspoon sea salt
- ½ teaspoon black pepper

TO PREPARE THE POT FISH:

Rinse fish carefully in ice water, drain, and pat dry with paper towels. Rub with lime juice. In a large skillet, heat oil over medium heat until hot, add the fish, and cook for about 2 to 3 minutes on both sides. Remove fish and set aside.

TO PREPARE THE STEW:

To the same pan, add the onion, garlic, ginger, cinnamon, thyme, and bay leaf. Sauté until onions are tender, about 2 minutes. Add seasoning pepper, tomatoes, wine, and 2 cups of water. Let it simmer for about 10 minutes. Stir in the bell pepper, half of the green onions and the okra, followed by the fish, one at time, to the stew. Simmer for about 5 minutes. Add cilantro and adjust for seasoning with salt and pepper.

TO SERVE:

Remove the pan from the heat. Spoon vegetables and fish onto a large platter. Top with remaining green onions.

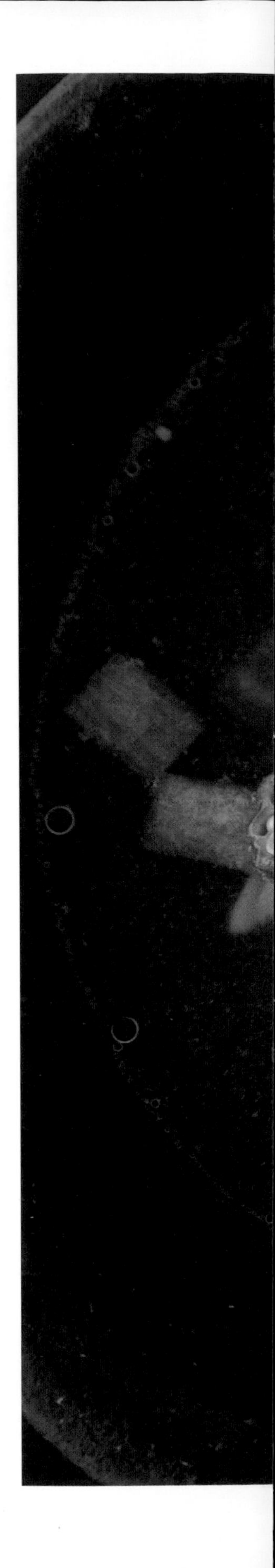

SERVES 4

CAST-IRON LIONFISH WITH PICKLED RED ONION

Pickling is used as a method of preserving fruits and vegetables on many Caribbean islands and throughout the world. The technique dates back centuries, to capture and infuse seasonal flavors. These pickled red onions add crunchy texture and a nice acidic contrast.

- 1 large red onion
- ¾ cup cider vinegar
- ¾ cup water
- 2 Tablespoons sugar
- 1 Tablespoon salt
- 1 teaspoon black peppercorns
- 1 teaspoon allspice berries
- 4 large lionfish fillets
- 1 teaspoon kosher salt
- ¼ teaspoon crushed red pepper flakes
- 2 Tablespoons coconut oil
- 2 cups kale, cabbage or swiss chard leaves, roughly chopped

TO PREPARE THE ONION:

Peel and slice the red onion into thin slices. Place into a pint-sized mason jar with a sealable lid.

TO PICKLE THE ONION:

In a small saucepan, stir together vinegar, water, sugar, and salt. Add in black peppercorns and allspice. Bring to a boil, then remove from heat. Pour mixture over onions in jar until full. Place the lid loosely on the jar and let cool to room temperature. Tighten the lid to seal and refrigerate. This will keep for several weeks refrigerated.

TO PREPARE THE LIONFISH:

Season the lionfish with salt and pepper flakes. In a large cast-iron skillet over moderate heat, heat the coconut oil. Add the lionfish and cook until nicely browned on both sides, about 2 minutes per side. Remove the fish from the pan. Using the same pan, wilt the kale quickly on high heat. Stir in 1 Tablespoon of pickle juice to the pan and cook for another minute.

TO FINISH:

Place the kale in the center of a large platter. Arrange the lionfish on the platter and garnish with pickled red onion.

Island recipes are handed down from one generation to another. Cast-iron pans are usually passed along too. A well-used heavy cast-iron pan carries deep flavors and many memories. I would describe that as a well broken-in pan. Technically, the cast iron maintains a consistent temperature and is great for browning fish and meats as well as for frying and stewing. I suggest that you start your own culinary cooking tradition with a good heavy cast-iron pan.

SERVES 4

BEACH SHACK SANDWICH

Delicious and easy to eat, fish sandwiches are adored throughout the Florida Keys and Caribbean. Grouper has been the fish of choice in Florida, and the demand for this popular fish has caused it to be over-fished. That has brought the need for seasonal catch regulations, making the classic grouper sandwich rare to find. Thinking sustainably, we should now be devouring lionfish sandwiches any chance we have.

- 4 large lionfish fillets
- 2 Tablespoons chopped cilantro
- 1 teaspoon sea salt
- ½ teaspoon crushed red pepper flakes
- 1 Tablespoon butter
- 1 Tablespoon pickled peppers with juice
- 4 4-inch sandwich buns, toasted
- 1 cup shredded savoy cabbage
- 1 large heirloom tomato, sliced

TO PAN FRY THE LIONFISH:

Season the fish with a Tablespoon of cilantro, salt, and red pepper. In a heavy pan over medium heat, melt the butter. When the butter is just about completely melted, add the lionfish, cooking the first side for 2 to 3 minutes. Turn the fish over and, with a spoon, glaze fish with the cooking liquid. Add the pickled peppers and continue to baste the fish until cooked through.

TO PREPARE THE SANDWICH:

Place a lionfish fillet into each toasted bun. Arrange the shredded cabbage and tomato on each bun. Drizzle with cooking juices and finish with remaining cilantro. Skewer with a bamboo toothpick.

In Saint Lucia, the most popular cooking pepper is the seasoning pepper. This small colorful pepper has a citrusy, aromatic warming flavor. To make pickled peppers, simply slice them thin and place in a sterile jar. Boil together a brine of cane vinegar, allspice, salt, and peppercorns. Pour the brine over the peppers in the jar. Let cool, cover, and refrigerate overnight.

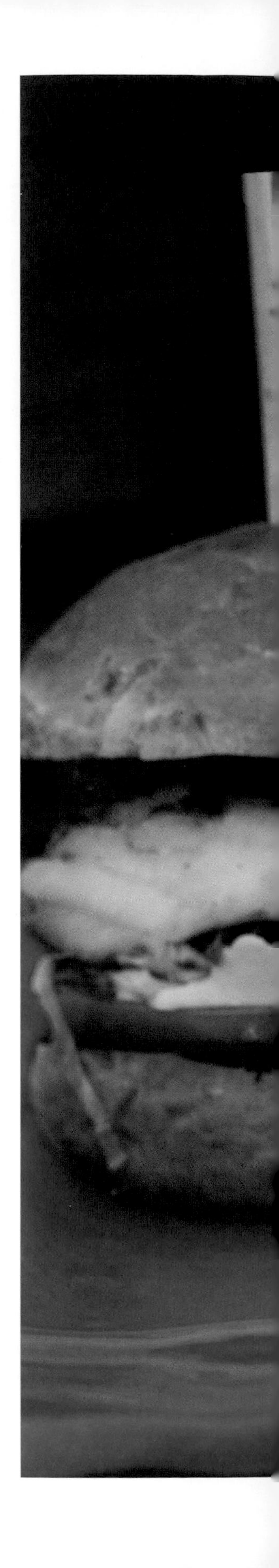

SERVES 4

LIONFISH ISLAND STYLE POKE SALAD

CHEF'S COLLABORATION: Sam Choy

Sam is the poke king! He goes to the fish market every day to find the freshest fish available for his pokes. He shared his secret for poke with us as he was in Japan enjoying a feast of fugu and sake! After all, the chefs that work with the usually poisonous fugu must be licensed in Japan.

- 4 Tablespoons light shoyu sauce
- 1 ½ teaspoons honey
- 1 ½ teaspoons garlic chili sauce
- 1 ½ teaspoons minced fresh bird chili pepper
- 2 Tablespoons ponzu lime juice
- 1 ½ pounds lionfish fillets
- ½ cup chopped green onions
- ½ medium red onion, sliced thin
- 2 cups mixed farmers greens
- 1 avocado, sliced
- 1 cup cherry tomatoes, cut in halves

TO PREPARE THE POKE SAUCE:

In a small bowl, mix together the shoyu, honey, garlic chili, bird chili, and ponzu. Cover and refrigerate to chill.

TO PREPARE THE POKE:

Rinse the lionfish under cold water and pat dry. Cut into half-inch pieces. Place lionfish in a large stainless steel bowl and add the sliced red onions and chopped green onions. Then pour the poke sauce into bowl, mixing gently, and chill.

And that is how you make fresh island poke.

TO SERVE:

In a colorful bowl, arrange the mixed greens and top with avocado and cherry tomatoes. Pour the poke mixture right on top of salad mix. Serve chilled and enjoy.

With eleven bestselling cookbooks, frequent Food Network appearances, multiple TV series, four James Beard Award nominations, a James Beard Foundation American Classics award, and global culinary recognition, Sam Choy is the world's go-to Hawaiian cuisine expert. He was crowned the "Godfather of Poke" as he opened Sam Choy's Poke-To-The-Max—a regional restaurant chain. Sam has been acclaimed as the cofounder and primary developer of Hawaiian Regional Cuisine and can be seen now on *Sam Choy in the Kitchen*.

SERVES 4

GRILLED LIONFISH WITH CURRY SAUCE

CHEF'S COLLABORATION: Cindy Pawlcyn

I love the ocean, and I love to eat sustainable seafood. We must continue to keep our oceans healthy for future generations and for our own survival.

For this dish, the curry sauce can be easily prepared a day ahead and reheated at serving time. Grill the fish on a hot barbeque or grill pan. Serve the lionfish with white or brown steamed rice and braised greens like collard or Jamaican callaloo.

Cindy Pawlcyn has pioneered fresh, seasonal, sustainable wine-country cuisine since opening Napa Valley's legendary Mustards Grill in 1983. Cindy has earned countless national awards and is a three-time James Beard Awards finalist for Outstanding Restaurateur. She is the author of five cookbooks, including the James Beard Award-winning *Mustards Grill Napa Valley Cookbook*. Cindy is a member of the Monterey Bay Aquarium Chefs Seafood Watch program.

- 3 Tablespoons peanut oil or other vegetable oil
- 1 medium onion, sliced into crescents, not rings
- 1 habanero chili pepper, seeded and sliced thinly or to taste
- 3 cloves garlic, sliced thinly
- 3 Tablespoons curry paste (Patak's Hot Curry Paste)
- 3 tomatoes, peeled and diced
- 1 cup white wine, fish stock, clam juice, or water
- 1 can coconut milk
- 5 Tablespoons butter
- 1 bunch collard greens or Jamaican callaloo
- 4 large lionfish fillets
- 1 teaspoon sea salt
- ½ teaspoon freshly ground pepper
- ½ teaspoon curry powder
- 4 pieces aluminum foil, cut into 12-inch squares
- 2 cups white or brown rice, cooked
- 1 large lime, cut in wedges
- 4 sprigs fresh cilantro leaves, stems removed
- 1 to 2 scallions thinly sliced or chives, chopped
- 2 Tablespoons vegetable oil

FOR THE CURRY SAUCE:

Over medium heat, heat the peanut oil in a skillet. When the oil is hot, add the onions and chili pepper and cook until tender and translucent, about 8 to 10 minutes stirring occasionally. Add the garlic and cook 1 to 2 minutes. Add the curry paste and stir until it is quite aromatic. Add the tomatoes, cooking 2 minutes, then add the wine. Bring the mixture to a boil, then reduce the heat to a low simmer, cooking until the sauce is jam-like. Add the coconut milk and 1 Tablespoon of butter to blend and heat. Set aside and reheat when serving.

FOR THE GREENS:

Strip the leaves from the stems of the collard greens or callaloo. Discard the stems. Finely slice the leaves. Heat a sauté pan over medium heat, add the butter and sauté until tender and the color is bright green. Season with salt and pepper to taste.

FOR THE LIONFISH:

Preheat the barbeque or grill pan. Bring the fish to room temperature and season liberally with salt, freshly ground pepper, and a dusting of curry powder on both sides.

TO COOK THE LIONFISH:

Liberally coat the center of each foil with the remaining butter. Place a fish fillet slightly below the center of the foil and fold the foil into a triangle, sealing the sides tightly. The sealed packet will look like a triangle. Repeat this with the other fillets and foil. Place the fish packets on the preheated hot barbeque or grill pan and cook for 3 minutes on each side, turning over once. Be careful when unwrapping the foil as the fish will have steamed and will be hot.

TO SERVE:

On a warm plate, add a half-cup of cooked rice, place a fish fillet on top of the rice and spoon a generous portion of the curry sauce over the fish. Add the collard greens to the side, a slice of lime and top the fish with a few cilantro leaves and scallions or chives.

SERVES 4

LIONFISH, WINGED BEAN "TABBOULEH" BLACK BEAUTY AJÍ DE AJO SAUCE

CHEF'S COLLABORATION: Robert Irvine

The interest for the winged bean as a legume has increased in past few decades. It is an exceptional legume and almost every part of the bean can be eaten. This plant is found abundantly in South Africa and Southeast Asia due to growing conditions and the plant's ability to avoid disease, such as in many areas of the Caribbean as it grows well in hot and humid countries.

Every part of the winged bean plant contains vitamin A, vitamin C, calcium, and iron among other nutrients. Being so rich in proteins, vitamins, minerals, and healthy fats, the pods are used in salad or in many vegetable dishes and side dishes.

BLACK BEAUTY AJÍ DE AJO SAUCE

- 1 each "black beauty" squash or dark green zucchini, split in half and scored
- 4 green ají de ajo chilies
- 1 each jalapeño
- 1 small white onion, peeled and cut in half
- ¼ cup grapeseed oil
- 2 teaspoons kosher salt
- 1 teaspoon ground black pepper
- 2 bunches cilantro, stems removed
- ¼ cup freshly squeezed lime juice, reserving zests

WINGED BEAN SALAD

- ½ cup shucked winged beans, blanched and chilled
- ½ cup finely diced English cucumber
- ½ cup finely diced tomato
- 1 cup thinly shaved kale
- ¼ cup finely diced red onion
- 4 mint leaves, shaved thin
- 2 Tablespoons finely chopped parsley
- 2 Tablespoons apple cider vinegar
- ¼ cup olive oil
- 1 bunch basil, shaved thin

LIONFISH

- 2 lemons
- ½ cup grapeseed oil
- 1 teaspoon kosher salt
- 4 large lionfish fillets
- 1 teaspoon olive oil
- 4 sprigs cilantro

TO PREPARE THE SAUCE:

The sauce should be made well ahead, even the day before since it's a cold sauce and it needs time to cool. Coat the black beauty squash, both types of peppers, and onions with the grapeseed oil. Add salt and pepper, then roast them under the broiler at 500 degrees. Every 3 to 5 minutes, turn them over so they roast evenly. Once they are all roasted and tender, let them cool briefly. In a stand-up blender, add the lime juice, lime zests, cilantro leaves, and all the roasted vegetables. Cover the blender and turn on high speed to puree the ingredients until smooth. Turn off the blender and taste for salt, pepper, and lime balance; add more if desired. Place in a small container and refrigerate until needed.

TO PREPARE THE SALAD:

Take each salad ingredient and mix together in a small bowl; add salt and pepper, and taste for balance. Add more salt, pepper, and cider vinegar if desired. Prepare this mix prior to the lionfish so it can rest in the refrigerator and build its flavors.

TO PREPARE THE LIONFISH:

In a small bowl, juice and zest one of the lemons. Add ⅔ of the grapeseed oil and ⅔ of the salt and mix together. Add the cleaned lionfish fillets and let marinate in the refrigerator for 15 to 30 minutes.

Remove the lionfish and lightly season with the remainder of the salt. On a stove top, bring a skillet to high heat and add the remainder of the oil to the pan. Gently lay the lionfish into the oil away from you to prevent splashing the hot oil, and pan sear for 1 to 2 minutes on each side to get a nice browning. When this is done, quickly remove, squeeze a little more lemon over the top, and serve immediately.

TO PRESENT THE DISH:

Remove the sauce and the salad from the refrigerator. To serve family style, spread the sauce decoratively on a large platter. Arrange the salad in the center of the platter. Place prepared lionfish over the salad. Drizzle a little olive oil over the lionfish and sauce, add some cilantro garnish, and serve.

Robert Irvine is an award-winning chef, fitness authority, and philanthropist best known for his long-running Food Network shows *Dinner: Impossible* and *Restaurant: Impossible*. Robert is the author of four books including his most recent, *Family Table*, which features simple and wholesome recipes for families. A tireless advocate of the military, Robert established The Robert Irvine Foundation in 2014 to support veterans and military causes.

SERVES 4

SMOKED LIONFISH SEA SALTINE NACHO

CHEF'S COLLABORATION: Stephen Phelps

As a source-conscious chef, I only allow sustainable seafood to enter my kitchen. As a fisherman as well, I see the effects that invasive species and environmental issues can have on our fisheries. Educating and preparing species people are unfamiliar with play important roles at Indigenous. Lionfish sells out every time we run it!

- 4 ounces smoked lionfish (or any sustainable smoked fish)
- 16 Sea Saltine Crackers, see recipe
- 1 cup Sauce Verde, see recipe
- 1 cup white cheddar, grated
- 3 Tablespoons micro cilantro
- 2 Tablespoons pickled jalapeños
- 2 Tablespoons pickled red onions
- 12 small heirloom tomatoes, sliced

SMOKED FISH BRINE

- 2 cups gluten-free soy sauce
- ½ cup spring water
- 2 Tablespoons brown sugar
- ¼ cup maple syrup
- 1 Tablespoon crushed red pepper flakes

SEA SALTINES CRACKER

- 2 cups unbleached flour
- 2 teaspoons sea salt
- ¾ cup spring water

SAUCE VERDE

- 1 poblano pepper, chopped
- 1 small onion, chopped
- 2 cups chopped tomatillos
- 2 cloves garlic, smashed
- ½ cup cilantro, chopped
- 2 limes, juiced
- ½ cup olive oil

TO PREPARE THE SMOKED LIONFISH:

Mix together the soy, water, brown sugar, maple syrup, and pepper flakes in a bowl and pour over fish in a ziplock bag. Refrigerate for at least 6 hours. Remove from brine and pat dry. Place in smoker for about 8 to 10 minutes on low. Finish in a 400-degree oven for another 5 to 10 minutes. Chill.

TO PREPARE THE CRACKERS:

Place flour in Cuisinart with salt. Slowly pulse in water until dough forms. Do not overmix. Roll out as thin as possible with dough pin. Using a fork or a docking tool, poke holes in the dough numerous times. Lay on parchment paper or silpat-lined sheet pan and bake at 300 degrees until golden and crisp. Allow to cool.

TO PREPARE THE VEGETABLES:

Toss vegetables in light amount of olive oil and pour onto a sheet pan. Bake at 400 degrees until they start to brown lightly. Remove and allow to cool.

TO PREPARE THE SAUCE VERDE:

Add to a blender the vegetables, cilantro, and lime juice and puree on low until blended. Slowly add in more olive oil until desired consistency.

TO FINISH AND SERVE:

Sprinkle Sea Saltine with grated white cheddar. Bake in the oven at 350 degrees until the cheese is melted. Arrange the diced smoked fish, jalapeños, onions, and tomatoes, and garnish with sauce and cilantro.

Chef Steve Phelps opened Indigenous in 2011 and since then has honed his reputation as one of the region's most inspirational and educational chefs, earning a James Beard Foundation semi-finalist nomination for Best Chef South. Chef Phelps is on the Monterey Bay Aquariums Seafood Watch Blue Ribbon Task Force and serves only sustainable seafood at Indigenous. Indigenous has helped to put Sarasota, FL, boldly on the map as a wonderful city in which to live, visit, and dine responsibly.

FALL

"Gourmandise is an impassioned, rational, and habitual preference for all objects that flatter the sense of taste."

—Brillat-Savarin

Wet Season

While weather patterns vary from island to island in the Caribbean, the one thing that is consistent about Caribbean weather is its year-round beauty. Trade winds bring steady sea breezes into the Caribbean from the northeast year-round. There is gorgeous weather with warm sunshine and the occasional tropical rain shower in fall. Some locals describe fall as bittersweet. After summer's heat and frantic pace, they welcome fall's slowdown. From grapefruit to starfruit to breadfruit, and from dasheen to pumpkins to sweet potatoes, autumn's bumper crop of fruits and vegetables offers a range of intense flavors and substantial textures.

Farmers markets are an exciting way to meet the Caribbean head on. Here you will see the ebb and flow of life on the island. It is a living cultural exchange that tells the story of its people and the vibe of the community. There are small street side markets almost every day and Saturday is market day on the waterfront in Saint Lucia's Soufriere with plenty of fish, home-grown vegetables, ground provisions, and hot peppers along with bunches of colorful herbs and aromatic spices.

There are numerous bustling open-air markets in the Caribbean. In Trinidad, the Chaguanas Market is a commotion of spices, fruits, and fish along with street foods like West Indian doubles, curry chicken roti, or bakes and shark. St. Georges Market Square in Grenada is full of the colorful aromas that ensconce this spice island, boasting freshly harvested and dried cinnamon bark, coriander, clove, mace, and nutmeg. Alongside these are locally grown tomatoes, soursop, callaloo leaves, and every kind of large and small local fish. Roseau New Market on Domenica is an old-time open market with flamboyant umbrellas covering equally multi-colored vegetables and catches of the day. Along the beach side in Guadeloupe is the Sainte Anne Market. The French and Creole flavors and ingredient varieties include local pumpkins, squashes, green leaves, and vegetables, as well as fresh tuna, mackerel, blue runner, and snapper.

A farmer's market is the best place to select your ingredients for this cookbook. Produce is fresh, ripe, and full of flavor with wholesome nutrients. Feel free to exchange and substitute for seasonal variables as you recreate these recipes.

JADE MOUNTAIN
Elijah Jules
Chef de Cuisine

Fall

SERVES 4

LIONFISH ESCABECHE

The world is constantly changing, and cuisine changes along with it. There's no better example of this than lionfish escabeche. With its cooking techniques rooted in the Mediterranean, this sweet and sour dish has evolved into a Caribbean Island classic. Other sustainable fish choices for making this escabeche include tuna, wahoo, tilefish, and swordfish.

- 8 lionfish fillets
- 1 teaspoon kosher salt
- 1 teaspoon freshly ground black pepper
- 2 Tablespoons chickpea flour
- ¼ cup olive oil
- 4 cloves garlic, thinly sliced
- 1 cup celery, peeled and thinly sliced
- 3 Tablespoons capers
- 3 Tablespoons cup raisins
- 3 Tablespoons small, green olives, pitted
- ¼ cup white wine vinegar
- ¼ cup water
- 1 large tangerine, segmented and juiced
- 2 Tablespoons fresh mint, chiffonade
- 2 Tablespoons cashews, toasted

TO PREPARE THE LIONFISH:

Season the lionfish with salt and pepper and dust with chickpea flour. In a large skillet over moderate high heat, heat the olive oil. Add the lionfish and cook until nicely browned on both sides, about 1 to 2 minutes. Remove the fish from the pan.

TO PREPARE THE VEGETABLES:

In the same pan, add the garlic and celery and cook about 4 minutes until golden. Add the capers, raisins, and olives, stirring until lightly glazed. Return the lionfish to the pan and add the vinegar, water, and tangerine juice. Cover and cook for 1 minute. Finish with the tangerine segments.

TO SERVE:

Spoon the vegetable mixture into the center of each plate, along with any remaining pan juices. Arrange the lionfish on top of this vegetable mixture. Sprinkle with fresh mint and garnish with pine nuts before serving.

CONCH AND LIONFISH FRITTER

The islanders love their fritters, whether Stamp and Go in Jamaica, Accras in Saint Lucia, Acras de Morue in Martinique, or Bacalaitos in Puerto Rico. This one brings together the rich conch flavor with the sweet taste of the lionfish.

- 1 cup buttermilk
- 4 large eggs
- 3 limes, juiced
- 1 teaspoon allspice
- 1 teaspoon cinnamon
- ½ teaspoon nutmeg
- ½ teaspoon clove
- 1 teaspoon salt
- ½ teaspoon black pepper
- 1 small onion, diced
- 3 stalks celery, diced
- 2 cloves garlic, minced
- 3 scallions, minced
- ½ pound conch, minced and tenderized with a mallet
- ½ pound lionfish, diced
- 2 cups all-purpose flour
- 2 Tablespoons baking powder
- 2 cups peanut oil
- ½ small Scotch bonnet pepper, minced

TO PREPARE THE FRITTER BATTER:

In a small stainless steel bowl, combine buttermilk, eggs, lime juice, and the spices, beat well, and set aside. In separate bowl, mix the onions, celery, garlic, scallions, conch, and lionfish. Fold in the flour and baking powder. Add wet ingredients to dry and lightly mix until just combined (lumps are okay).

TO COOK THE FRITTERS:

Preheat peanut oil in a deep saucepan to 375 degrees. Spoon Tablespoon sized scoops of fritter mixture into the hot oil, about ten per batch. Take care to not overcrowd the pan. Allow to fry to golden brown and remove to a paper-lined plate.

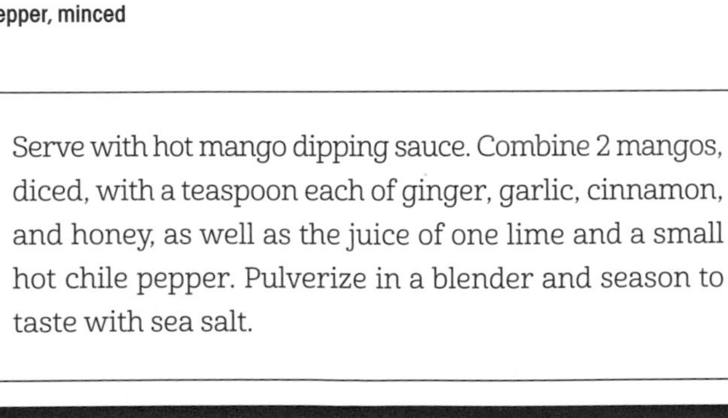
Serve with hot mango dipping sauce. Combine 2 mangos, diced, with a teaspoon each of ginger, garlic, cinnamon, and honey, as well as the juice of one lime and a small hot chile pepper. Pulverize in a blender and season to taste with sea salt.

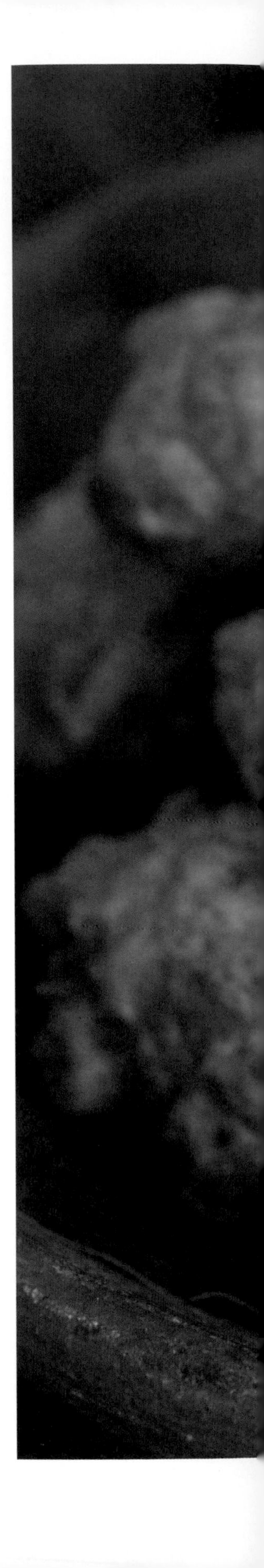

SERVES 4

LIONFISH AND SAKE CLAMS

A small quantity of sake will enhance the flavor of this dish. The alcohol in the sake evaporates while the food is cooking, and only the flavor remains. Sake (SAH-keh, not saki) is made from rice and water. The quality of sake depends on the quality of rice and water being used for brewing.

- 2 Tablespoons unsalted butter
- 4 large lionfish fillets
- 1 Tablespoon minced garlic
- 1 large shallot, julienned
- ½ cup finely diced red bell pepper
- 2 medium bananas, peeled and cut into rounds
- 32 small Littleneck clams, washed under cold running water
- ¼ cup low-sodium soy sauce
- ½ cup sake
- ½ bunch cilantro, chopped plus 2 sprigs for garnish
- ½ bunch scallions, chopped
- ½ teaspoon sea salt
- ½ teaspoon freshly ground black pepper

TO PREPARE THE LIONFISH:

Place a sauté pan over medium heat. Add 1 Tablespoon of butter to melt. Place the lionfish into the hot melted butter and brown the first side. Remove the lionfish from the pan to a warm holding pan.

TO PREPARE THE CLAMS:

In the first pan, add another Tablespoon of butter and stir in the garlic, shallot, bell peppers, and banana. Cook until fragrant, then add the clams, soy sauce, and sake. Cover and simmer for 2 minutes, then carefully place the partially cooked lionfish back into the pan on top of the vegetables to steam until the clams open.

TO SERVE:

Arrange the cooked lionfish, clams, and vegetables in large bowls. Toss the cilantro and scallions into the cooking broth, and pour it over the bowls. Whole cilantro sprigs to garnish.

SERVES 4

SIZZLE AND SPICE LIONFISH WITH LOCAL PINEAPPLE

Tropical fruits and big spices are a large part of island cooking. Fresh local Saint Lucian pineapples are grown in rich volcanic soil with warm tropical rainfall and sunshine.

- 1 teaspoon salt
- 2 large limes, juiced
- 2 teaspoons soy sauce
- 1 teaspoon sugar
- 1 Tablespoon ginger, minced
- 4 4-ounce lionfish fillets
- 4 Tablespoons extra-virgin olive oil
- 2 teaspoons freshly chopped cilantro
- ¼ cup fresh orange juice
- 2 teaspoons lemon juice
- 1 teaspoon kosher salt
- ½ teaspoon coarse ground black pepper
- ½ ripe pineapple, cut small wedges
- 4 small Roma tomatoes, cut in half inch pieces
- 1 cup farmer's greens
- 2 teaspoons black sesame seeds

TO PREPARE THE LIONFISH:

In a small stainless steel bowl, mix salt, lime juice, soy, and ginger. Pour marinade over fish in a shallow dish and marinate for 1 hour. Strain and dry the lionfish with toweling paper before cooking.

TO COOK THE LIONFISH:

In a heavy bottom grill pan over high heat, drizzle 1 Tablespoon of olive oil. Place the lionfish in pan and cook for 2 to 3 minutes on each side until golden brown.

TO PREPARE THE SALAD:

In a small bowl, whisk together the cilantro, orange juice, lemon juice, remaining olive oil, salt, and pepper. Separately combine the pineapple, tomatoes, and farmer's greens. Toss them with the vinaigrette.

TO SERVE:

Arrange the pineapple and salad on a platter. Place the lionfish in the center. Finish the salad with black sesame seeds.

SERVES 4

FENNEL CRUSTED LIONFISH WITH SWEET POTATO HASH

Over time, East Indian influences found their way into West Indian cooking. In the ancient thinking of the Ayurveda, fennel is called "samf" or "saunf," and is believed to taste bitter, pungent, and sweet. It is considered neutral to slightly warming, energetic, and balancing. Fennel is nourishing to the brain and calming to the spirit.

- 2 teaspoons fennel seeds
- 1 teaspoon whole white peppercorns
- 1 teaspoon dill
- 1 teaspoon thyme
- ½ teaspoon salt
- 4 large lionfish fillets
- 3 Tablespoons olive oil
- 2 Tablespoons dry white wine
- 2 cups sweet potato hash

SWEET POTATO HASH

- 2 medium sweet potatoes
- 3 Tablespoons olive oil
- 2 large shallots, julienned
- 2 cloves garlic, chopped
- 2 pieces star anise, ground
- ½ cup diced sweet red peppers
- 1 Tablespoon diced serrano chili
- 1 Tablespoon chopped cilantro
- 1 Tablespoon kosher salt

TO PREPARE THE LIONFISH:

In a dry pan over medium heat, toast the fennel seeds and whole white peppercorns until aromatic, approximately 1 to 2 minutes. Crush the spices. Use these along with dill, thyme, and salt to season the lionfish well on both sides. Drizzle with 1 Tablespoon of olive oil and set aside until ready to cook.

TO COOK THE LIONFISH:

Heat a large non-stick pan on medium-high heat. Add the remaining olive oil. Sauté the lionfish on the first side, browning it well for 3 minutes. Then turn it over. Cook for another 3 minutes until tender. Add the white wine and remove the fish from the heat to fish plates. Arrange the sweet potato decoratively on the plate.

TO PREPARE THE SWEET POTATO:

Preheat the oven to 350 degrees. Wash and dry the sweet potatoes. Place the potatoes directly on the oven rack and bake them for 45 minutes. They should be slightly soft to the touch. Remove from the oven and let them cool completely. Peel the sweet potato and dice into large half inch pieces.

TO PREPARE THE HASH:

In a large, flat cast-iron pan, warm the olive oil on medium heat. Add the shallots and cook until caramelized, approximately 3 to 4 minutes. Add the garlic and the diced sweet potato. Toss together and cook for another minute. Then add the star anise, sweet red peppers, serrano chili, and cilantro. Sauté together for another minute and season with salt.

SERVES 4

LIONFISH AND LOBSTER BOUILLABAISSE

Caribbean Island cooking is filled with plenty of delicious fish soups, fish teas, and fish pepper pots. This is a fish-based soup laden with lobster, fish, and chunks of delicious goodness in every bite.

- 2 Tablespoons olive oil
- 1 cup diced onion
- ½ cup diced celery
- ½ cup diced carrot
- 1 teaspoon salt
- 1 whole seasoning pepper
- 1 vanilla bean, split open
- 1 quart fish stock
- 1 cup white wine
- 1 cup coconut milk
- 1 cup chayote, diced
- 2 small green bananas, sliced
- 2 Tablespoons coconut oil
- 2 spiny lobster tail, with the shell on, cut each in 4 to 5 pieces
- 4 small lionfish fillets
- 4 small kingfish steaks
- 2 sprigs fresh tarragon

TO PREPARE THE FISH BROTH:

In a large pot, heat the olive oil. Add the onions, celery, carrot, and ginger. Season with salt and seasoning pepper. Sauté for about 2 minutes. Add the vanilla bean along with fish stock and wine. Bring the liquid to a boil and reduce to very slow simmer. Stir in the coconut milk, chayote, and banana. Cook for 15 minutes. Remove from the heat, keeping warm.

TO PREPARE THE BOUILLABAISSE:

Using a heavy pot with a lid, warm the coconut oil and place the lobster pieces in and cook over medium-high heat for 3 to 4 minutes until the lobster shell turn rosy pink. Carefully ladle in the coconut broth. Add the fish fillets to the pot. Slowly heat up to a soft simmer. Cook until the fish and the lobster are tender.

TO SERVE:

Evenly divide the lobster and fish fillets in 4 deep bowls. Ladle in the bouillabaisse and vegetables. Serve garnished with fresh tarragon.

Lobster season varies from island to island in the Caribbean and changes sometimes even year to year. The season is scheduled by sustainable resource managers working to protect the spiny lobster from being over-fished. In Saint Lucia, the lobster season generally starts in August and runs through March.

SERVES 4

WOOD ROASTED LIONFISH WITH ROOT VEGETABLES

Ground provisions is the term used in West Indian nations to describe several traditional root vegetable staples such as yams, sweet potatoes, dasheen root (taro), eddos, and cassava. They are often cooked and served as a side dish in local cuisine.

- 2 carrots
- 1 dasheen
- 2 small yams
- 1 bulb fennel
- 2 leeks, white only
- 2 Tablespoons olive oil
- 8 small lionfish fillets
- 2 shallots, minced
- ¼ cup cognac
- 1 cup dry white wine
- 2 Tablespoons butter, room temperature
- 2 Tablespoons parsley, minced
- 2 Tablespoons chives, minced
- ½ teaspoon of sea salt

TO PREPARE:

Cut vegetables into natural two-inch shapes, and partially cook (blanch for 3 minutes) in boiling water. Start with carrots, then dasheen, yams, fennel, and leeks.

TO WOOD ROAST:

Heat oil in a large cast-iron pan and sauté lionfish, making sure to sear the first side. Turn the fish and add blanched vegetables, and place in wood fire oven for five minutes. Remove pan from oven and, before returning to stovetop, remove the cooked lionfish. Add shallots and cognac and flambé. Add white wine and reduce to ¼ liquid. Season lionfish with sea salt.

TO SERVE:

Place lionfish and vegetables decoratively on platter. Return sauce to stove, slowly adding soft butter and fresh herbs. Pour sauce over lionfish and vegetables on platter and serve.

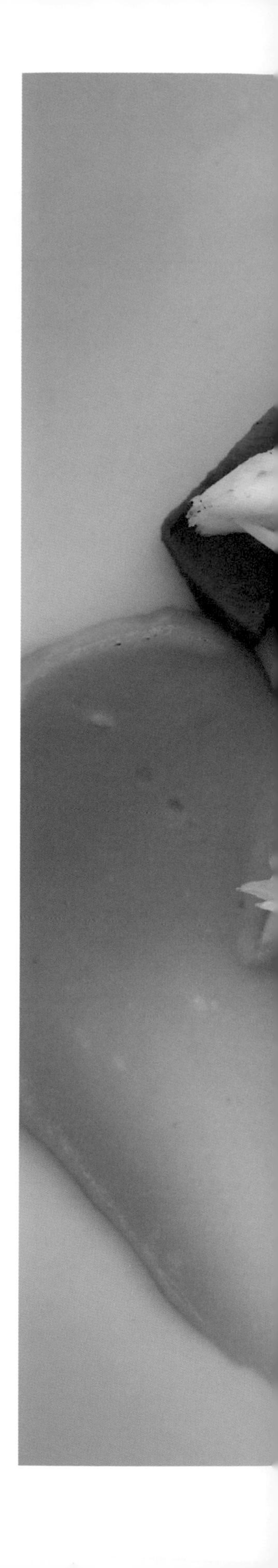

SERVES 4

SEA SALT ROASTED LIONFISH WITH WEST INDIAN SQUASH

Calabaza is a type of pumpkin-like squash that is round and varies in size. It can be as large as a watermelon or as small as a cantaloupe. The color of calabaza can also vary and may include greens, tans, reds, and oranges. Some squash is all one color while other calabaza are multi-colored and may include all of colors listed above.

- 1 teaspoon cinnamon
- ½ teaspoon fennel seeds
- ¼ teaspoon nutmeg
- 1 teaspoon fresh thyme leaves
- 1 teaspoon sea salt
- 1 teaspoon freshly ground black pepper
- 1 large West Indian squash, cut in large cubes
- 4 large lionfish fillets
- 3 Tablespoons olive oil
- 2 Tablespoons pomegranate molasses
- ½ teaspoon Maldon sea salt
- 1 large lime, cut in wedges

BLEND THE SPICES:
In a small bowl, mix the cinnamon, fennel, nutmeg, thyme, salt and pepper. This will be used to season both the fish and squash. Brush both fish and squash with olive oil.

TO PREPARE THE SQUASH AND LIONFISH:
Preheat the oven to 375 degrees. In a large roasting pan place the squash in single layer and roast for 15 minutes until soft. Add the fish to the roasting pan. Brush the cooked squash lightly with pomegranate molasses and continue roast together for another 5 minutes.

TO SERVE:
Arrange the lionfish and squash on a platter. Sprinkle with Maldon salt and garnish with lime segments.

SERVES 4

PAN FRIED LIONFISH WITH BITTERS SPICED CHICKPEAS

Bitters are essential to a well-crafted cocktail today. Why not use them in thoughtful cooking? After all, bitters are simply an herbal concoction made from roots, barks, and spices including cardamom, nutmeg, and cinnamon. These are classics in the Caribbean spice box.

- 1 cup cooked chickpeas
- ½ teaspoon ground cumin
- ½ teaspoon smoked paprika
- ¼ teaspoon ground ginger
- 1 teaspoon kosher salt
- ½ teaspoon ground black pepper
- 1 Tablespoon olive oil
- 1 Tablespoon orange bitters
- 2 teaspoons freshly chopped parsley
- 2 Tablespoons butter, softened
- 4 large lionfish fillets
- 1 Tablespoon rice wine vinegar
- 1 Tablespoon chopped scallions
- 1 small lime, cut in wheels

TO PREPARE CHICKPEAS:

Preheat the oven to 400 degrees. Mix the chickpeas with cumin, paprika, ginger, and ¼ teaspoon each of salt and pepper. Drizzle in the olive oil. Using a low ovenproof pan, roast chickpeas for 15 minutes, shaking the pan occasionally. Add bitters and continue to roast for another 5 minutes until golden brown and crunchy. These can be made ahead.

TO PAN FRIED THE LIONFISH:

Season the fish with parsley and the remaining salt and pepper. In a heavy pan over medium heat, melt the butter. When the butter has just about completely melted, add the lionfish, cooking the first side for 2 to 3 minutes. Turn the fish over and with a spoon glaze the fish with the cooking liquid. Add the rice wine vinegar and continue to baste the fish until cooked through.

TO SERVE:

Arrange the lionfish on a plate. Toss the spiced chickpeas into the remaining cooking liquid from the fish along with the scallions. Spoon the chickpeas over the fish. Garnish with fresh lime wheels.

SERVES 4

LIONFISH WITH STEWED FARMER'S VEGETABLES

Farmer's vegetables can be a host of choices for this stew. Use what is seasonal and available at your local farmers market. The dasheen is a starchy root vegetable used throughout the Caribbean. You can choose sweet potato or calabaza as a delicious substitute.

- 1 teaspoon kosher salt
- 1 teaspoon freshly ground black pepper
- 1 ½ teaspoons curry powder
- 4 large lionfish fillets
- ¼ cup olive oil
- 2 cloves garlic, thinly sliced
- ½ cup onion, peeled and thinly sliced
- 1 teaspoon freshly grated ginger
- ½ cup dasheen, peeled and diced
- ½ cup orange juice
- ½ cup zucchini, thinly sliced
- ½ cup ripe tomato, seeded and diced
- 2 Tablespoons small Kalamata olives, pitted
- 2 Tablespoons fresh mint, torn
- 2 Tablespoons fresh cilantro, picked

TO PREPARE THE LIONFISH:

In a small bowl combine the salt, pepper and curry. Season the lionfish with half of the mixture, reserving some for the vegetable stew. In a large skillet over moderate high heat, warm the olive oil. Add the lionfish and cook until nicely browned on both sides, about 2 minutes. Remove the fish from the pan and keep warm.

TO PREPARE THE VEGETABLES:

In the same pan, add the garlic and onion, cooking about 4 minutes until soft. Add the ginger and dasheen, stirring in the orange juice. Bring to a boil and lower to a simmer before adding in zucchini and tomato. Simmer for 10 minutes until thickened and season with remaining curry mixture.

TO SERVE:

Spoon the vegetable mixture into the center of each plate along with any remaining pan juices. Arrange the lionfish on top of this vegetable mixture. Sprinkle with fresh mint and cilantro before serving.

SERVES 4

BANANA LEAF GRILLED LIONFISH

Banana leaves have been used for centuries in cooking throughout the tropical world. The leaf protects the flesh from burning while allowing it to cook in its own juices and enhances the fish with an aromatic sweet essence.

- 4 sprigs fresh cilantro
- 4 sprigs fresh flat leaf parsley
- 2 sprigs fresh mint
- 1 thumb-sized piece of freshly peeled ginger
- 1 small red Fresno chile, seeded
- 1 teaspoon kosher salt
- ¼ teaspoon ground allspice
- ¼ teaspoon ground cumin
- ¼ teaspoon ground cinnamon
- 1 ½ Tablespoons cane vinegar
- 4 Tablespoons peanut oil
- 4 large pieces of fresh or frozen banana leaves, approximately 10 inches square
- 4 large lionfish fillets
- 10 pieces hearts of palm
- 1 large lime, cut in wedges

TO PREPARE THE SPICE PASTE:

In a blender, combine the leaves of the cilantro, parsley, and mint along with the ginger, chile, salt, allspice, cumin, cinnamon, and vinegar. Pulse together until a smooth paste forms. With the blender running on low speed, drizzle in the oil until incorporated. Pour mixture into a small bowl and allow flavors to mellow together for at least 30 minutes.

TO PREPARE THE LIONFISH:

On a large clean work area, arrange the banana leaves. Using a pastry brush, paint the center of each leaf with spice paste. Place the lionfish in the center and brush the fish liberally with the spice paste. Proceed to fold each leaf bundle like a small package and secure with a bamboo toothpick.

TO GRILL THE FISH:

Preheat the grill to medium heat. A wood fire grill will add the most flavor, but a gas grill will do the job. Grill the banana leaf wrapped lionfish for about 4 to 5 minutes on each side. At the same time, grill the hearts of palm, brushing each with the remaining spice paste.

SERVES 4

ORZO SAUTÉ WITH LIONFISH

CHEF'S COLLABORATION: Susur Lee

Growing up in Hong Kong surrounded by water, cooking and eating fresh fish and seafood is second nature to me. I grew up sharing whole steamed fish with my family, and the smell of a fish market always brings me back home. Sustainable is always the healthier option, and I find it nourishes my body and mind as well.

Praised as one of the "Ten Chef's of the Millennium" by *Food & Wine* magazine, Susur Lee is still at the top of his game. Between helming four restaurants in Toronto—Lee, Luckee, Kid Lee, and Lee Kitchen—and overseeing his prestigious TungLok Heen in Singapore, Chef Lee makes numerous television appearances (*Iron Chef America*, *Chopped Canada*, *Top Chef Masters*, *MasterChef Asia*, and *Iron Chef Canada*) and travels the globe as guest chef and consultant. He was recently named an ambassador to Canada's 150th birthday celebrations and was awarded the Lifetime Achievement Award by Canada's 100 Best. Balancing the epicurean traditions of China with the classical techniques of French cuisines, Chef Lee improvises a daring and original culinary aesthetic.

- 4 large lionfish fillets
- 2 cups orzo pasta
- ½ cup vegetable oil
- 2 thinly sliced shallots
- 2 eggs
- 2 green onions
- 3 Tablespoons brown butter
- ½ teaspoon finely ground garlic
- ½ teaspoon finely ground ginger
- 2 cups loosely packed julienned spinach
- 2 Tablespoons pine nuts, toasted
- Soy Mirin Sauce
- ½ cup mirin
- ¼ cup soy sauce
- 2 Tablespoons water
- 1 Tablespoon bonito flakes

TO PREPARE THE SOY MIRIN SAUCE:

In small pot, combine mirin, soy sauce, and water. Heat on medium-low until liquid boils. Remove from heat and add dried bonito flakes. Cover and store in refrigerator for two days. Strain through fine mesh sieve.

TO PREPARE THE LIONFISH:

Place the fillets in a steamer basket over simmering hot water. Cover and steam for 3 to 4 minutes. Remove, break apart, and set aside.

TO PREPARE THE ORZO:

In a medium sized pot, boil orzo pasta in unsalted water until cooked al dente. Strain.

TO CRISP THE SHALLOTS:

In a medium sauce pot, heat oil to 325 degrees. Add shallots and fry until crispy and golden brown. With a slotted spoon, remove shallots from oil and drain on a paper towel.

FOR THE OMELETS:

Whisk eggs with 2 Tablespoons of vegetable oil. Heat a non-stick pan on medium-high. Add just enough of the egg mixture to make a thin layer on the bottom of the pan. Cook until set. Carefully remove omelet from pan. Repeat three more times. Cut each omelet into a 4-inch square, the trimmings can be julienned.

TO PREPARE THE GREEN ONIONS:

Wash green onion under cold running water. Slice the white part of the green onions into rings and julienne the green parts. Set aside.

TO BRING THE FISH TOGETHER:

In a medium pan, add brown butter, garlic, and ginger. Cook until golden brown. Add cooked orzo pasta, broken up lionfish meat, and green onion rings. Sauté until warmed throughout. Pour in 4 to 5 Tablespoons of prepared soy mirin sauce. Remove from heat and stir in spinach.

TO FINISH AND SERVE:

Arrange the orzo and fish mixture on a large white platter. Garnish with pine nuts, crispy fried shallots, omelet, and julienne green onion. Serve immediately.

SERVES 4

PAN ROASTED LIONFISH, SQUID INK FIDEOS, CHORIZO EMULSION

CHEF'S COLLABORATION: Éric Ripert

At Le Bernardin, we spend our days evaluating seafood, which means not simply the quality, but also the ethics and sustainability of how it is harvested. We believe deeply in supporting the artisanal fishers that are seeking out the most sustainable methods and species. What we've recently discovered is that lionfish, an extremely invasive and detrimental species, are not difficult to catch and in fact have delicious and very versatile flesh. They are truly one of the best examples of sustainable fishing that I'm happy to encourage!

SQUID INK SAUCE

- 1 Tablespoon canola oil
- 1 Tablespoon garlic, minced
- 3 Tablespoons shallot, minced
- 4 Tablespoons mussel stock
- 1 Tablespoon squid ink

FIDEOS

- 2 ounces capellini pasta
- 1 Tablespoon garlic, sliced thin
- 3 Tablespoons shallot, sliced thin
- 3 Tablespoons red bell peppers, sliced
- 3 Tablespoons yellow bell pepper, sliced
- ½ cup white wine
- ½ cup chicken jus
- 1 cup mussel stock

TO MAKE THE SQUID INK SAUCE:

In a small amount of canola oil, sweat the garlic and shallot in a sauce pot until soft. Add the mussel stock, and simmer for 5 to 7 minutes. Add the squid ink and cook for 2 to 3 minutes more. Pass through a chinois, and reserve.

TO PREP THE FIDEOS:

Place capellini pasta on a baking sheet, and toast in a 350 degree oven until dark golden brown, or about 5 minutes. Break the pasta into one-inch pieces. In a sauce pot, sweat the garlic, shallots, and red and yellow peppers till soft, then deglaze with white wine and reduce by ¾. Add chicken jus and mussel stock, and simmer for 8 to 10 minutes. Strain the liquid through a chinois into a new sauce pot. Cook the pasta in the liquid until al dente; reserve.

GARLIC CHIPS

- 1 large clove elephant garlic
- 1 teaspoon pimento Fennel Radish Salad
- ½ head fennel, shaved thin on mandolin, stored in cold water
- 2 red radishes, sliced then cut into matchsticks
- 4 pieces of citrus lace, cut in half
- 1 scallion green, cut into small julienne
- 1 teaspoon olive oil
- ½ lemon

CHORIZO EMULSION

- 1 cup sweet chorizo
- 1 cup spicy chorizo
- 1 Tablespoon garlic, sliced thin
- 3 Tablespoons shallots, sliced thin
- 3 Tablespoons white wine
- ½ cup mussel stock
- ½ lemon
- 2 Tablespoons Canola oil

TO MAKE THE GARLIC CHIPS:

Peel garlic and slice very thinly on a mandolin. Blanch garlic in milk for 2 to 3 minutes, and dry on a cloth. Fill a sauce pot halfway with canola oil, and place over medium-high heat. Add the blanched garlic to the oil, and, with chopsticks, stir gently to keep the garlic from sticking. After 2 to 3 minutes, the garlic will begin to take on color. When it is a light shade of golden brown, remove the chips with a slotted spoon and drain excess oil on paper towels. Season lightly with fine sea salt and pimentón, and reserve.

TO PREPARE THE CHORIZO EMULSION:

Peel the outer skin off of the chorizo and cut into small pieces. In a sauce pot, add enough canola oil to coat the bottom half-inch of the pot. Add the chopped chorizo, and cook over low heat for 45 minutes, stirring occasionally to render all the fat out. Strain the oil through a chinois and reserve both the oil and rendered chorizo. In a second sauce pot, sweat the garlic and shallots in canola oil until soft, then add the rendered chorizo and deglaze the pan with white wine. Add the mussel stock and bring to a simmer. Strain the infusion through a chinois into a clean sauce pot, and, using a hand blender, slowly emulsify the reserved chorizo oil into the infused mussel stock. Season with fine sea salt and lemon juice, and reserve.

LIONFISH

- 4 large lionfish fillets
- 2 Tablespoons Wondra flour, for dusting
- Fine sea salt and freshly ground pepper

TO COOK THE FISH:
Season the lionfish with fine sea salt and freshly ground pepper, then dust with Wondra flour on one side of the fillet. Sauté the fish Wondra-side down in a medium-hot, heavy-bottomed pan until a crust appears, then place the pan into a 400 degree oven and cook until desired doneness. Remove the fish from the pan, and let it rest. While the fish is resting, warm the fideos in a small sauce pot, adding just enough squid ink sauce to absorb into the pasta and keep it moist. Mix the fennel, radishes, citrus lace, and scallion with olive oil and a squeeze of lemon juice. Season with fine sea salt and freshly ground pepper.

TO PLATE:
Use a slotted spoon to drain the fideos and arrange them neatly in the center of the plate. Slice the lionfish into 1.5 cm thick slices, and shingle five slices in a straight line over the pasta. Garnish the top of each serving with the fennel and radish salad, and two garlic chips. Sauce is served hot and poured around the fish just before serving.

Éric Ripert is the chef and co-owner of Le Bernardin, New York's internationally acclaimed four-star seafood restaurant, which holds 3 Michelin Stars, is ranked No. 26 on the World's 50 Best Restaurant list, and most recently was named No. 1 worldwide by La Liste. Additionally, Chef Ripert is the author of five cookbooks plus his *New York Times* bestselling memoir *32 Yolks: From My Mother's Table to Working the Line*. Chef Ripert is the vice chairman of the board of City Harvest, New York's first food rescue organization fighting hunger in NYC.

SERVES 4

WOK TOSSED SALT AND PEPPER LIONFISH

CHEF'S COLLABORATION: Andrew Zimmern

I love to cook Chinese food. And these crispy, savory, spicy pieces of lionfish are the perfect way to enjoy this delicious fish. Serve with steamed Chinese rice and a nice plate of sautéed Chinese greens cooked simply with ginger and scallion. I make this recipe once a week in my house and everyone loves it. Recently, my son requested this dipping sauce be served with it, a favorite from another fried seafood dish and culinary perfection was achieved.

- 1 pound small lionfish fillets
- 2 Tablespoons kosher salt
- 1 teaspoon ground white pepper
- 1 teaspoon sea salt
- 1 Tablespoon sugar
- 2 cups peanut oil
- 2 large egg whites
- ¼ cup corn starch
- 6 scallions, cut in 2-inch lengths
- 1 large dried red chile, diced

DIPPING SAUCE

- ½ cup shallots, sliced paper thin
- ⅓ cup good, naturally brewed soy sauce
- ¼ cup rich homemade chicken broth
- 2 Tablespoons fresh lime juice
- 1 Tablespoon brown sugar
- ½ cup fresh cilantro leaves, chopped
- 2 serrano chiles, sliced thin with the seeds

TO PREPARE THE FISH SEASONING:

Toss the lionfish with kosher salt. Cover and let sit for 15 minutes. Separately combine the sea salt, pepper, and sugar. Reserve seasoning for later.

TO COOK THE LIONFISH:

Rinse the fish under cold water and press with a dry white kitchen towel to remove moisture. Heat peanut oil to 375 degrees over high heat in a wok. Dip the fish into the egg white, then dredge in corn starch. Fry to crisp golden brown in 2 batches. Drain off all but 1 Tablespoon of oil from the wok and increase the heat. When it begins to smoke, add the scallions, chile, and crispy fish fillets. Toss and actively season with spice mixture, tossing as you go.

TO SERVE:

Arrange on a large platter and serve dipping sauce on the side.

TO PREPARE THE DIPPING SAUCE:

Combine the shallots, soy sauce, broth, lime juice, and brown sugar. Refrigerate for 2 hours. Mix well and stir in the cilantro and chiles.

A four-time James Beard Award-winning TV personality, chef, author, and teacher, Andrew is regarded as one of the most versatile and knowledgeable personalities in the food world. As the creator, executive producer, and host of Travel Channel's *Bizarre Foods* franchise, *Andrew Zimmern's Driven by Food*, *The Zimmern List*, and Food Network's *Big Food Truck Tip*, he has explored culture through food in more than 170 countries.

WINTER

"Cookery is not chemistry. It is an art.
It requires instinct and taste rather than exact measurements."
—James Beard

Festive Season

When it comes to weather, the Caribbean is renowned for having arguably the most desirable climate on the planet. It's never cold, never roasting, and rain falls to a predictable rhythm—keeping these tropical islands lush and leafy. The best time to visit the Caribbean is generally considered to be festive season, when it's slightly cooler, drier, and less humid, and tourists flock to escape the northern winter. Winter is the festive season in all manners of the word. Festivities begin as locals and visitors enjoy the Christmas holidays and New Year's celebrations on through to many traditional island fetes ending with Fat Tuesday, which signals the conclusion of Carnival and the beginning of Lent.

Winter's evening cool breezes and clear blue skies that reflect the sea water bring out another amazing story of the local bounty, including parrotfish, blue runner, kingfish, snapper, conch, sea urchin, and spiny lobster. Across the Caribbean, seafood reigns supreme in local fare. However, with varied cultural backgrounds and regional preferences, different islands lay claim to different flavors and specialty cuisines, making local food festivals events that any hungry visitor is going to want to take a bite of!

This dry season produces rooty yams, dasheen, sweet potatoes and cassava—which are often referred to as ground provisions. You can find these boiled, roasted, and stewed as dumplings or vegetables. Festive season is carambola, golden apple, Barbados cherries, hearty winter eggplants, squashes, sweet peppers, and hot peppers. This is citrus season too in the Caribbean; that means numerous varieties of sweet and sour oranges, limes, grapefruits, and tangerines. Our winter produce offers a surprising range, thus creating a Caribbean fusion of flavors.

Get inspired by these recipes.

SeaLife

Winter

SERVES 4

LIONFISH CLUB SLIDERS

A slider is an American term for a small sandwich, typically around two inches across, made with a bun. This zesty lionfish is enhanced with crispy bacon and deliciously served on a toasted bun.

- 2 ribs celery, peeled and minced
- 1 large shallot, minced
- ½ teaspoon lemon zest
- 1 Tablespoon red wine vinegar
- ½ teaspoon Dijon mustard
- 2 teaspoons fresh squeezed lime juice
- 2 Tablespoons mayonnaise
- 1 pound lionfish fillet, cooked and flaked
- ½ cup lump crab meat
- 1 Tablespoon chopped cilantro
- 1 Tablespoon chopped capers
- 1 teaspoon sea salt
- 8 2-inch Brioche rolls
- 1 Tablespoon softened butter
- 8 strips crisp Applewood smoked bacon
- 2 small ripe Roma tomatoes
- 1 cup farmers salad greens

TO PREPARE THE LIONFISH SALAD MIXTURE:

In a stainless steel bowl, mix the celery, shallot, lemon zest, red wine vinegar, mustard, lime juice, and mayonnaise. Add the lionfish, crabmeat, cilantro, and capers. Season with sea salt. Refrigerate salad for at least 1 hour.

TO PREPARE THE CLUB SLIDER:

Brush the buns with butter and toast. Arrange the bacon slices, tomato, and greens on each bun. Scoop the lionfish mixture into each bun. Skewer with a bamboo toothpick.

For a variation on the theme, this same salad mixture can be used to make mini fish cakes by shaping and coating them with crushed cracker meal. Then pan fry them carefully in melted butter until golden brown. Serve in the same fashion on toasted buns with crispy bacon and tomatoes.

SERVES 4

SMOKED LIONFISH DIP

We smoke our lionfish over charcoal that is made from local sustainable hard woods. This gives this smoked fish its unique Saint Lucian flavor. This dip can be served with crisp plantain chips dusted with sea salt and cayenne.

- ½ cup mayonnaise
- ⅓ cup coconut milk
- 2 Tablespoons sour cream
- 3 small scallions, chopped
- 1 teaspoon freshly diced jalapeño
- 2 Tablespoons freshly squeezed lime juice
- 6 Tablespoons freshly chopped cilantro
- 16 ounces smoked lionfish
- 1 teaspoon kosher salt
- ¼ teaspoon cayenne
- 1 Tablespoon toasted sesame seeds

TO PREPARE THE DIP:
Using a small bowl, blend the mayonnaise, coconut milk, sour cream, scallions, jalapeño, lime juice, and cilantro until smooth. Fold in the smoked lionfish and season to taste with salt and cayenne.

TO SERVE:
Chill dip well for 2 hours. Serve in a freshly cracked coconut shell, top with toasted sesame seeds.

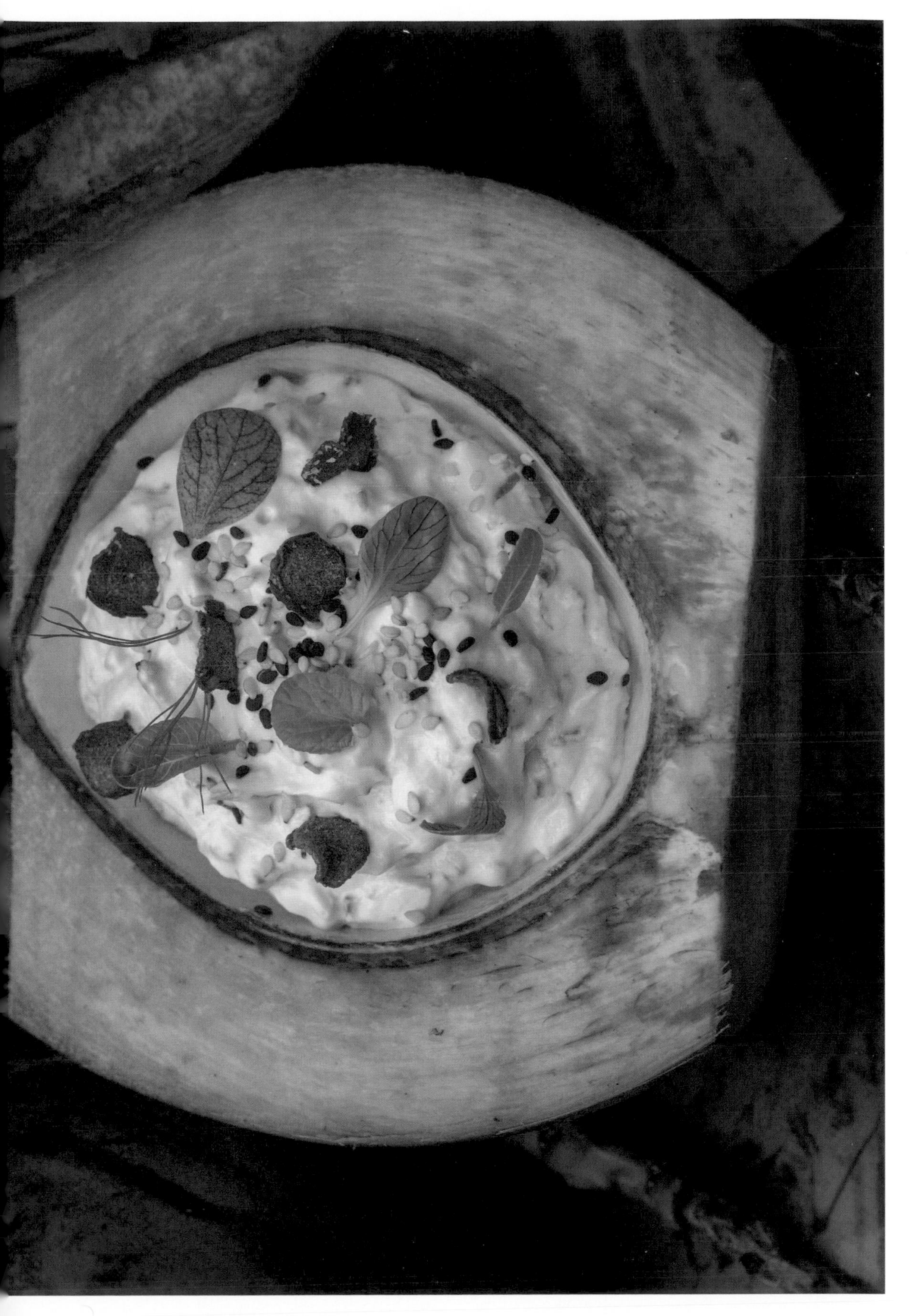

SERVES 4

CREOLE LIONFISH WITH SEASONING PEPPERS

A Saint Lucian favorite, seasoning pepper is such a practical name for these little guys, as that is precisely what they do. It's true: they look a lot like the habanero or Scotch bonnet pepper. But these don't even come close to the heat level of one of those peppers and truly season food with an incredible flavor. These peppers are grown throughout the Caribbean and are known formally as ají dulce peppers. The flavor is straight habanero without the heat—that tangy richness and sweetness and just a slight touch of heat.

- 8 small lionfish fillets
- 1 teaspoon kosher salt
- ½ teaspoon freshly ground black pepper
- ½ teaspoon ground cumin
- 1 lime, juiced
- 3 seasoning peppers, diced
- 1 red pepper, diced
- 1 green pepper, diced
- 1 onion, diced
- 1 celery stalk, diced
- 2 cloves garlic, minced
- 2 sprigs fresh thyme
- 1 bay leaf
- 2 teaspoons tomato paste
- 1 cup dry white sauvignon blanc wine
- 1 teaspoon orange bitters
- 1 vine ripe tomato, diced
- 1 Tablespoon freshly chopped cilantro

TO SEASON THE LIONFISH:

Season the lionfish with salt, pepper, cumin and lime juice. Cover and allow to cure for 30 minutes.

TO PREPARE THE VEGETABLES:

Heat olive oil in medium saucepan. Add seasoning peppers, peppers, onions, garlic, thyme, and bay leaf, then tomato paste. Simmer vegetables for 5 to 6 minutes stirring occasionally.

TO PREPARE THE LIONFISH:

Add lionfish and wine into the hot vegetable pot and allow to cook for 3 minutes. Stir in diced tomato, bitters, and season with salt and pepper to taste. Simmer for 5 to 6 minutes until flavor blend.

TO SERVE:

Carefully spoon the vegetables into a platter and top with the lionfish. Garnish with chopped cilantro.

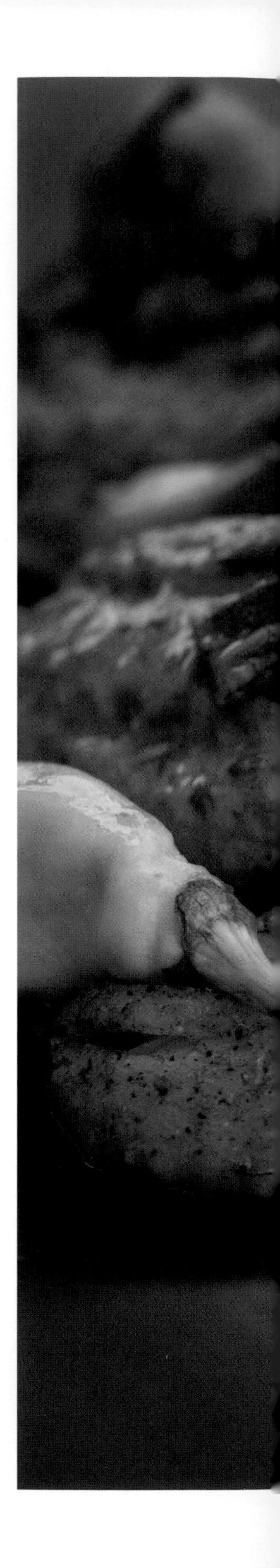

SERVES 4

CURRIED CARIBBEAN LIONFISH

Caribbean curry powder has a more aromatic flavor than Madras curry powder as it has more turmeric and often includes allspice. What makes this dish so enticing is the coconut milk coupled with Scotch bonnet pepper. Cilantro spiked rice would be a delicious accompaniment.

- 4 medium scallions, finely chopped
- 1 teaspoon fresh thyme, chopped
- 2 cloves garlic, finely chopped
- ¼ teaspoon minced Scotch bonnet pepper
- ½ teaspoon Angostura bitters
- 2 large limes, juiced
- 4 large lionfish fillets
- 4 Tablespoons olive oil
- 1 large sweet onion, diced
- 1 Tablespoon Caribbean curry powder
- 1 cup coconut milk
- 1 teaspoon coarse sea salt
- ½ teaspoon freshly ground black pepper
- 2 Tablespoons fresh chopped cilantro

TO MARINATE THE LIONFISH:

In a medium-sized stainless steel bowl, combine the scallions, thyme, garlic, scotch bonnet, and bitters with the lime juice. Marinate the lionfish in this mixture for 30 minutes.

TO PREPARE THE CURRY:

Using a heavy bottomed cast-iron skillet, warm 2 Tablespoons of olive oil until just before it begins to smoke and add in the onions. Cook over medium heat until they begin to caramelize, then stir in the curry powder. Cook for another 2 minutes and push the mixture to one side of the pan.

TO COOK THE LIONFISH:

Add the remaining oil while turning the heat up to high. Carefully place the marinated fish into the hot oil. Pan fry the lionfish on the first side for about 2 minutes, turning the fish with a spatula when golden brown. Add the remainder of the marinade along with the coconut milk and bring to a simmer. Season with salt and pepper.

TO SERVE:

When the fish is cooked through, remove to a serving plate. Stir the sauce well to reduce another minute and finish with fresh cilantro. Spoon the sauce over the lionfish.

SERVES 4

CHANDON BENI LIONFISH AND LOCAL HONEY ROASTED ORGANIC CARROTS

Chandon beni is a key ingredient in Saint Lucian cooking. The herb is used mostly in the Southern Caribbean. Yet, it is notorious in Puerto Rician cooking where it is called culantro. Chadon beni is very similar in flavor to cilantro (coriander) but considerably fuller in flavor and pungency.

- 4 large lionfish fillets
- 1 Tablespoon chopped chandon beni
- 2 teaspoons kosher salt
- 1 large lime, cut in 6 wedges
- 2 Tablespoons extra-virgin olive oil
- ½ pound organic baby carrots, multi-colored with stems
- 3 sprigs fresh thyme
- 1 teaspoon cumin seed
- ½ teaspoon crushed red pepper flakes
- 1 Tablespoon local honey

TO PREPARE THE LIONFISH:

Season the lionfish with chandon beni, a teaspoon of salt, and the juice of 2 lime wedges. Set aside covered for 30 minutes. Drizzle with 1 Tablespoon extra-virgin olive oil before roasting.

TO PREPARE THE ORGANIC CARROTS:

Toss the carrots with remaining olive oil, thyme, cumin, pepper flakes, and remaining salt.

TO ROAST TOGETHER:

Preheat the oven to 400 degrees. Place the carrots in an ovenproof pan large enough for both the carrots and fish. Start with the carrots and roast for 5 to 6 minutes. Add the honey to glaze the carrots. Push the carrots to one side, allowing enough room for the fish to be placed in the pan. Lower the temperature to 350 degrees and continue to roast fish and carrots together for 4 to 5 minutes.

TO SERVE:

Place the lionfish on a large white platter with the carrots arranged decoratively. Garnish with lime wedges.

SERVES 4

LIME, GARLIC, AND CILANTRO LIONFISH

This recipe cracks the code of Caribbean cooking. After grasping this humble dish, you will have the confidence to cook Caribbean. The unpretentious combination of fresh ingredients and simple technique will result in a briny lionfish appetizingly balanced by the sweet richness of coconut and citrus notes with a bright herbal finish.

- 8 small lionfish fillets
- 1 teaspoon kosher salt
- 1 teaspoon freshly ground black pepper
- 3 Tablespoons coconut oil
- 2 cloves garlic, thinly shaved
- 1 Tablespoon seasoning peppers, chopped
- 1 large lime, cut in 6 wedges
- 2 Tablespoons fresh cilantro leaves

CREOLE SPICE

- 1 teaspoon ginger
- 1 teaspoon coriander
- 1 teaspoon allspice
- ½ teaspoon cinnamon
- 1 teaspoon cumin
- ½ teaspoon crushed red pepper flakes
- 2 teaspoons kosher salt

TO PREPARE THE LIONFISH:

Season the lionfish with salt and pepper. In a large cast-iron skillet over moderate high heat, heat 2 Tablespoons of coconut oil. Add the lionfish and cook until nicely browned on both sides, about 1 to 2 minutes. Add the garlic and seasoning peppers. Squeeze 2 wedges of lime over the fish. Add half of the cilantro. Remove the fish from the pan.

TO SERVE:

Arrange the lionfish in the center of a plate. Spoon pan juices over the fish. Sprinkle with remaining fresh cilantro, drizzle with the reserved coconut oil, and garnish with lime wedges before serving.

Spice it up using one of the Caribbean spice box seasonings instead of salt and pepper.

SERVES 4

GRILLED LIONFISH WITH AVOCADO SALSA VERDE

Fresh, authentic salsa verde will always add something special to your cooking. Salsa verde can be prepared in many ways—it can be made with roasted or raw tomatillos and can be combined with many different types of herbs. We add avocado to ours for a richer flavor. Without a doubt, this is a salsa that, once you've tried it, you won't want to live without it.

- 6 medium tomatillos, charred lightly over very high heat in a cast-iron pan
- 2 medium avocados, peeled and sliced
- 1 small jalapeño, seeded
- 2 cloves garlic, peeled
- ½ bunch cilantro leaves, picked, leaving a few leaves for garnish
- 1 large lime, juiced
- 1 teaspoon sea salt
- 8 small lionfish fillets
- 2 Tablespoons olive oil

TO PREPARE THE SALSA VERDE:

Place tomatillos, 1 avocado, jalapeño, garlic, cilantro, lime juice, and salt into a blender and pulse twice and then let run for 30 seconds continuously. Pour into a mason jar, cover, and refrigerate for 1 hour before serving.

TO PREPARE THE LIONFISH:

Preheat the grill. Season the lionfish with salt and pepper. Brush the olive oil on the fish and on the grill. Grill the lionfish for 2 to 3 minutes on each side until golden brown.

TO SERVE:

Arrange the lionfish on a colorful platter. Decoratively add the remaining sliced avocado and cilantro. Serve with salsa verde.

The trick to grilling fish is to preheat the grill and brush it clean. You should also carefully rub the grates with cooking oil just before placing the lightly oiled fish on the grill. If you have removed the skin from the fish, then the fleshy side should be grilled first. Maintain good heat and when the edges of the fish start to brown it is time to carefully flip the fish with a spatula. Other good choices of fish you can grill with this recipe are grouper, snapper, and cobia.

SERVES 4

SESAME SEED CRUSTED LIONFISH WITH TANGERINES

With a nutty crunch and sweet flavor, sesame seeds are a popular ingredient used extensively in East Indian and West Indian cooking.

- 1 Tablespoon sesame seed
- 1 teaspoon cumin seed
- 1 teaspoon lemon zest, minced
- ½ teaspoon kosher salt
- ½ teaspoon coarse ground black pepper
- 8 small lionfish fillets
- 2 Tablespoons olive oil
- 6 medium tangerines, 4 juiced and 2 sliced for garnish
- 1 sprig thyme
- 6 each peppercorns
- 2 Tablespoons shallots, minced
- 3 Tablespoons rice wine vinegar
- 6 Tablespoons butter, at room temperature

TO PREPARE THE LIONFISH:

In a bowl, combine the sesame seeds, cumin seeds, lemon zest, salt, and pepper. Roll the lionfish in the seed mixture, pressing to adhere. Drizzle with olive oil and set aside until ready to cook.

TO PREPARE SAUCE:

In a small sauce pot, combine tangerine juice, thyme, peppercorns, shallots and vinegar. Simmer for 10 minutes or until reduced by three quarters. Whisk in the butter off the heat while still warm in thumb sized pieces. Whisk until sauce thickens and the butter is emulsified into the sauce. Strain through a fine sieve and reserve sauce just warm.

TO COOK THE LIONFISH:

Preheat a cast-iron pan hot. Drizzle in olive oil and sear the lionfish for 1 minute on each side, cooking it crisp.

TO SERVE:

Pour the tangerine sauce on the plate and place the crisp lionfish on top. Garnish with slices of tangerine.

SERVES 4

GRILLED LIONFISH WITH SHRIMP AND RUM

Rum is produced on almost every island in the Caribbean. It plays a role in the culture of most of the islands in the West Indies. I suggest using a tasty rum like Bounty from Saint Lucia or Mount Gay from Barbados. The flavor and quality of the rum will shine through the seafood.

- ½ cup lime juice, freshly squeezed
- ¾ cup amber honey
- ½ cup dark rum
- 2 teaspoons fresh ginger root, grated
- 1 clove garlic, chopped fine
- ¼ teaspoon cayenne
- 2 Tablespoons butter
- 12 large shrimp, peeled and deveined with the tail end of the shell left on
- 4 large lionfish fillets
- 1 teaspoon sea salt
- ½ teaspoon black pepper, freshly ground
- 1 Tablespoon olive oil
- ¼ cup cilantro, finely chopped
- 1 large lime, cut into wedges

TO PREPARE THE SAUCE:

Combine the lime juice, honey, rum, ginger, garlic, and cayenne in a saucepan. Stir until thoroughly blended. Bring sauce to a simmer over medium heat, cooking for 5 minutes. Whisk in cold butter piece by piece until slightly thickened. Do not boil. Set aside and keep warm.

TO PREPARE THE LIONFISH:

Preheat the grill. Season both the lionfish and the shrimp with salt and pepper. Drizzle with olive oil. Grill the fish and shrimp over medium heat carefully until the fish is done on both sides. Continue to grill the shrimp until nicely browned.

TO SERVE:

Stir the cilantro into the sauce. Spoon 2 Tablespoons of sauce over the grilled shrimp and toss carefully. Spoon the remaining sauce onto the serving platter. Arrange the lionfish with the shrimp decoratively. Finish with lime wedges.

SERVES 2

CRISPY CURRY FRIED WHOLE LIONFISH

The lionfish spines—which can be venomous—must be removed very carefully. I mention it here because you are using a whole lionfish. The fish itself should be cleaned and rinsed in cold water. A fresh mixture of herbs and spices is essential for flavor.

- ½ cup cilantro
- ½ cup parsley
- ½ cup scallion
- 2 cloves garlic
- 1 teaspoon allspice
- 2 teaspoons Caribbean curry powder
- 1 teaspoon brown sugar
- 1 large lime, juiced
- ¼ cup olive oil
- 1 large whole lionfish
- 1 quart vegetable oil for frying
- ½ cup flour

TO PREPARE GREEN SEASONING SAUCE:
In a blender, pulse together the cilantro, parsley, scallion, garlic, allspice, curry, salt, sugar, and lime juice until pureed together. Drizzle in the olive oil with the blender running. Using a rubber spatula, pour the sauce into a glass container with a lid.

TO PREPARE THE LIONFISH:
Cut three slashes in the thickest part of the lionfish on both sides. In a small bowl, mix 3 Tablespoons of cold water with 3 Tablespoons of the green seasoning. Use this to marinate the fish in a large ziplock bag for at least 30 minutes and up to 2 hours.

TO FRY THE LIONFISH:
Remove the fish from marinade and shake off excess liquid. Place flour in a shallow plate. In a pan large enough hold the fish, heat the oil to 325 degrees. Once oil is hot, coat fish in flour and gently slide coated fish into hot oil and fry 5 minutes on each side, carefully turning the fish until golden brown in color. When the fish is evenly golden all over, remove and drain on paper towel.

TO SERVE:
Serve hot with additional green seasoning sauce on the side.

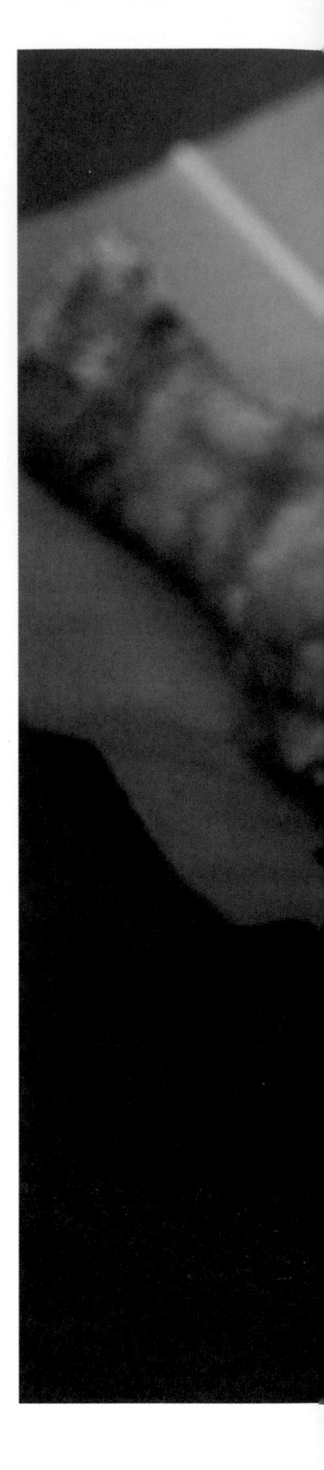

Whole fish cookery is very popular in the Caribbean. Probably most small fish or pot fish are served this way in homes and in local restaurants. Many locals will fight over the fleshy sweet fish cheeks. Regretfully, not everyone is willing to do the work to pick out the tastiest morsels of fish.

SERVES 4

POACHED LIONFISH WITH PEAS AND RICE

Peas and rice are a very traditional dish in the Caribbean. Each region has its own twist on the recipe. In Jamaica, red kidney beans are used. Cubans use black beans and Grenadines use pigeon peas as do the people of Puerto Rico.

- 2 Tablespoons vegetable oil
- 1 medium onion, minced
- 1 clove garlic, minced
- 1 large seasoning pepper, seeded and diced
- 1 cup uncooked Basmati rice
- 2 teaspoons kosher salt
- 1 teaspoon freshly ground black pepper
- 1 cup pigeon peas, cooked
- 1 can unsweetened coconut milk
- 5 sprigs fresh thyme
- 1 whole bay leaf
- 4 large lionfish fillets
- 2 scallions, minced
- 2 cloves

TO PREPARE THE RICE AND PEAS:

Heat the oil in a small heavy pot and brown the onion along with the garlic and seasoning pepper. Stir in the rice, 1 teaspoon salt and some of the pepper until well coated. Pour in the pigeon peas, one cup of the coconut milk, and 1 cup of water along with a sprig of thyme and bay leaf. Bring to a boil and lower heat to a simmer covering to cook for 20 minutes over medium-low heat. Remove from heat and fluff the rice with a fork. Allow to stand covered for another 5 minutes.

TO POACH THE LIONFISH:

Season the lionfish with remaining salt and pepper. Place the fillets in a heavy skillet just large enough to hold the lionfish. Add the scallion and cloves to the pan and pour the remaining coconut milk into the pan along with enough cold water to just cover the fish. Cover with parchment paper cut to the size of the pan. Over very low heat slowly warm the coconut milk broth to a simmer. Allow to simmer for 2 to 3 minutes depending on the thickness of the fillets. When just done in the center of the fillets, remove from heat and let the fish rest in the broth.

TO SERVE:

Arrange the rice, peas, and poached fish on a large colorful plate. Garnish with thyme sprigs.

SERVES 4

KEY'S LIONFISH CAKES WITH LEMON CAPER MAYONNAISE

CHEF'S COLLABORATION: Andre Bienvenu

For more than a hundred years, no visit to Miami has been complete without stopping in at Joe's Stone Crab. From the beginning, it has always been the love of food, family, and friends that has brought in customers and kept them coming. Stone Crabs are seasonal and sustainable yet rare. Market prices are negotiated at the docks with the fisherman based on availability. Florida law forbids the landing of whole Stone Crabs. Fishermen can take claws at least 2 ¾ inches long and are required to return Stone Crabs safely to the water. The Stone Crab can regenerate its claws three to four times, making it a renewable resource and a favorite among conservationists and foodies.

Andre's personal kitchen philosophy is that learning should never stop. For the past twenty years, Andre has been the executive chef of Joe's Stone Crab in Miami Beach, the world renown 104-year classic restaurant. In addition, Andre is a leading community-minded chef and has been a featured guest chef at Pebble Beach Food & Wine, LA Food & Wine, and South Beach Wine & Food festivals. Andre served as chef chairman for Share Our Strength's No Kid Hungry Taste of the Nation-Miami. He has Food Network appearances, local TV appearances, and a James Beard Award nomination.

LIONFISH CAKES

- 1 pound fully cooked, steamed lionfish
- 2 Tablespoons chopped shallots
- 4 Tablespoons fine diced red peppers
- 4 Tablespoons fine diced green peppers
- 1 Tablespoon chopped garlic
- 2 Tablespoons fresh lemon juice
- ¼ teaspoon Worcestershire sauce
- ⅛ teaspoon ground black pepper
- ⅛ teaspoon onion powder
- ¼ teaspoon garlic powder
- 1 teaspoon kosher salt
- ¼ teaspoon old bay seasoning
- ½ cup mayonnaise
- 1 egg yolk
- 5 Tablespoons panko breadcrumbs
- 4 Tablespoons vegetable oil
- 1 lemon, cut in wedges

LEMON CAPER MAYONNAISE

- 1 cup mayonnaise
- 1 teaspoon capers
- 4 Tablespoons fine diced green tomato
- 1 Tablespoon chopped parsley
- 2 Tablespoons fine diced red onion
- ½ teaspoon kosher salt
- ¼ teaspoon ground black pepper
- ¼ teaspoon old bay seasoning

TO PREPARE THE LIONFISH CAKE:

In a large bowl mix together the scallions, peppers, garlic, shallots, lemon juice, Worcestershire sauce, and dry spices. Fold in the mayonnaise and egg yolk. Carefully flake apart the lionfish and add to the vegetable mix. Add panko until the mixture binds together. Form the mixture into 3-inch disks and refrigerate.

TO PREPARE LEMON MAYONNAISE:

n a small bowl, mix the mayonnaise, capers, tomato, parsley, onion, salt, pepper, and seasoning. Cover and refrigerate to chill.

TO COOK THE CAKES:

In a heavy cast-iron pan, warm the oil until it just begins to smoke. Lower the heat in the pan and, in small batches, pan fry the lionfish cakes until golden brown, about 2 minutes on each side.

TO SERVE:

Arrange the lionfish cakes on a colorful platter. Serve with lemon mayonnaise and fresh lemon wedges.

SERVES 4

SIZZLING TANGERINE MISO LIONFISH KEBABS

CHEF'S COLLABORATION: Tom Douglas

I love the drama of serving kebabs on a sizzle platter of charred and caramelized onions. Try using a cast-iron pan and bring it directly to the table while it's still super-hot. Pour a little mirin and sake over the top so the pan sizzles, sputters, and smokes. Yipee!

TANGERINE MISO TERIYAKI GLAZE

- 1 cup fresh tangerine juice
- ½ cup mirin
- ¼ cup soy sauce
- 2 Tablespoons red miso
- 2 Tablespoons firmly packed brown sugar or palm sugar
- 2 teaspoons peeled and grated fresh ginger
- ½ teaspoon chopped garlic
- ½ teaspoon grated tangerine zest
- 1 teaspoon cornstarch plus 1 teaspoon water

LIONFISH

- 1 ½ pounds (24 ounces) lionfish fillets, cut into one-inch pieces
- 8 skewers, preferably double-pronged skewers which keep the chunks anchored better
- Kosher or sea salt and freshly ground black pepper as needed
- The sizzle platter
- A cast-iron pan
- 3 Tablespoons canola oil, or as needed
- 1 cup sliced red onion
- 1 cup sliced yellow onion
- 8 scallions, trimmed and sliced
- Zest from 3 lemons, coarsely grated (on a coarse grater)

TO FINISH THE KEBABS

- Lemon wedges
- Coarse sea salt
- ½ ounce mirin plus ½ ounce sake mixed together

Often described as the chef who put Seattle on the culinary map, Tom has earned national recognition, including the James Beard Award for Outstanding Restaurateur. Tom is the author of four cookbooks and cohosts a weekly radio show called Seattle Kitchen. Tom Douglas's restaurants include Dahlia Lounge, Etta's, Seatown, Rub with Love Shack, Palace Kitchen, Lola, Serious Pie, Serious Biscuit, Dahlia Bakery, Trattoria Cuoco, Brave Horse Tavern, Assembly Hall Juice and Coffee, TanakaSan, and Home Remedy.

TO MAKE THE TERIYAKI GLAZE:
Combine the tangerine juice, mirin, soy sauce, miso, brown sugar, ginger, garlic, and zest in a small saucepan over medium heat. Simmer and reduce by half, about 10 minutes. Make a slurry by mixing the cornstarch with the water. Add the slurry to the simmering glaze and allow to simmer another minute. The glaze should be as thick as maple syrup. Set aside. (Note: if you make the glaze ahead and chill it, you will need to warm it up and whisk it before brushing it on the skewers.)

TO PREPARE THE LIONFISH:
Thread the lionfish pieces on the skewers, dividing up the pieces of fish evenly among the 8 skewers. (Use about 3 ounces of fish pieces on each skewer.) Season the fish with salt and pepper. Fire up the grill. Grill the skewers over direct heat, turning as needed and brushing with glaze as they cook until they are cooked through, about 5 minutes total time. The sugar in the glaze can burn, so watch carefully and move the skewers to a cooler part of the grill if needed. (Note: It doesn't take long to grill the fish, so you may want to have a helper setting up the "sizzle platter" at about the same time you start grilling.)

TO SET UP THE SIZZLE PLATTER:
Put a cast-iron pan over high heat. Add the oil. When the oil is hot, add the red and yellow onions, the scallions, and lemon zest and cook a few minutes until the onion mixture is slightly charred and caramelized, stirring as needed. Remove the pan from the heat and top with the fish kebabs, right off the grill. Season the fish with a squeeze of lemon and a sprinkle of coarse finishing salt. Immediately, while the pan is still sizzling hot, place it on a trivet right on the table in front of your guests and drizzle the mirin-sake mixture over the kebabs. The platter should be hot enough so the mirin-sake sizzles dramatically and aromatically. Serve immediately. (Note: Use a potholder to hold the handle of the pan. Don't put the pan directly on the table; use a trivet or other heatproof protection for the table. Warn guests that the pan is hot!)

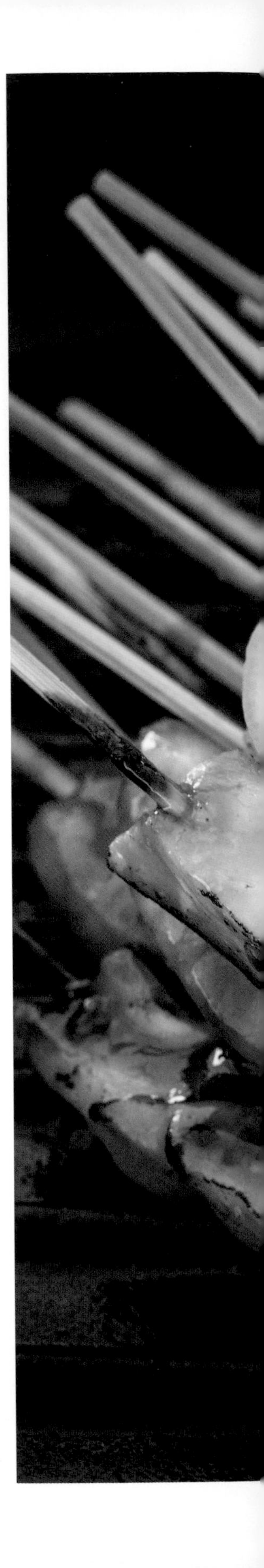

SERVES 4

RICE FLAKED LIONFISH WITH DRY GINGER BROTH

CHEF'S COLLABORATION: Floyd Cardoz

This is one of my favorite ways to cook fish. It utilizes rice flakes as a crust and gives the fish texture. The rice flakes mimic the scales of the fish. The rice flake crust helps prevent the fish from falling apart and makes cooking a breeze. The sun-dried ginger makes an amazingly light and flavorful broth.w

Chef Floyd Cardoz is a celebrated Indian American chef in New York City, India, and throughout the world. Floyd is the culinary director of two restaurants in Mumbai, India, The Bombay Canteen and O Pedro. He runs The Bombay Bread Bar, a casual modern Indian restaurant in Soho, NY. Cardoz is the author of *One Spice, Two Spice* and *Floyd Cardoz: Flavorwalla*. Cardoz is a four-time James Beard Award nominee. Floyd won season three of Bravo's *Top Chef Masters*.

- 4 cups fish stock or chicken stock
- 2 roma or plum tomatoes, quartered
- 2 teaspoons canola oil
- 2 cloves
- ¼ teaspoon kalonji (available at Indian spice stores)
- ¼ teaspoon cumin, whole
- 2 cloves fresh garlic, peeled and sliced
- 1 teaspoon fresh chopped ginger
- ½ medium white onion, peeled, quartered, and sliced thin
- ¼ teaspoon turmeric
- ½ teaspoon ground dry ginger
- ¼ fresh green chilies, deseeded and sliced thin
- ½ bunch cilantro, leaves picked, and stems tied in a bouquet
- 1 teaspoon kosher salt
- ½ teaspoon freshly ground black pepper

LIONFISH

- 4 large lionfish fillets
- 1 egg
- 2 Tablespoons all-purpose flour
- 1½ cups of rice flakes (called Poha, available at Indian spice stores)
- Salt
- Freshly ground black pepper
- ½ cup canola oil

TO PREPARE THE STOCK:

In a medium size pot, bring the stock and tomatoes to a boil; simmer over medium heat until reduced by half, approximately 10 minutes. Pass through China cap or a fine mesh strainer without pressing too much of the solids through.

TO PREPARE THE DRY GINGER BROTH:

Place a sauce pot over medium heat add the oil and the whole spices. Cook for one minute and add the sliced garlic. Cook slowly until golden and add the ginger and the onion cook until transparent. Add the turmeric, dry ginger and chilies. Cook for one more minute and then add the reduced tomato-stock reduction. Simmer for 5 minutes add the salt and infuse with the cilantro stems. This broth can be made up to a day in advance.

TO PREPARE THE LIONFISH:

In a bowl, combine the egg and the all-purpose flour and mix to a smooth paste. Using a brush, paint the egg wash on the skin side. Place the fillets in the rice flakes until they form a layer. Season the fillets on both sides with salt and pepper. Heat a skillet over medium-high heat, then add the oil. When hot, add the fish (rice flake side down) and cook for 2 to 3 minutes or until golden brown. Turn over in the pan and cook for an additional 1 to 2 minutes.

TO SERVE:

Arrange the lionfish with the crispy side up on a platter and garnish with cilantro leaves.

SERVES 4

SMOKED LIONFISH HADDIES WITH CRISPY POTATO, PRESERVED CABBAGES, AND DILL

CHEF'S COLLABORATION: Jonathon Sawyer

As a young Clevelander, fishing out on the Great Lake of Erie was one of the most indelible memories of my childhood. Those first yellow perch, steelhead, and walleye pike we caught, killed, gutted, and cooked would forever shape my respect for fish, and food in general. After all, the majority of the known and unknown of planet Earth is ocean, and as chefs and stewards of our planet's dining culture it's up to us to preserve what we have. Eat more sustainably harvested fish.

- 2 Tablespoons cultured salted butter
- 1 teaspoon cracked, toasted caraway
- ¼ cup sauerkraut
- ¼ cup kim chee
- 2 cups peeled, grated, starchy potatoes (Russet, Kennebec or Yukon)
- 1 teaspoon potato starch
- 3 Tablespoons fresh picked dill
- 4 large lionfish fillets, hot smoked
- 1 Tablespoon extra-virgin olive oil
- 1 lemon, zested and wedged
- 1 Tablespoon crème fraiche
- 1 Tablespoon whole grain mustard
- ½ ounce sustainable caviar
- 1 Tablespoon malt vinegar
- 1 teaspoon salt
- ½ teaspoon freshly ground black pepper

TO PREPARE THE CABBAGE:

Heat a small saucepan over medium heat, and add half of the butter to the pan. Once the butter is melted, add the caraway and allow spice to bloom for 30 seconds. Add cabbages to the saucepan, cover, and slowly cook for 10 minutes stirring occasionally.

TO READY THE POTATO:

In a large sauce pot, bring aggressively salted water to boil. Add grated peeled potato and blanch for approximately 3 minutes until potato is fully cooked through. Carefully strain off water in a colander. Season gently with salt and pepper.

TO PREPARE THE PANCAKE MIXTURE:

Fold the cabbage mixture into the potato; the mix should be starchy and sticky to the touch. If it is not, add a dash of potato starch. Add half of the dill to the mix and allow to cool, then add half of the lionfish and leave the mixture in the refrigerator covered in the colander. This can be done a day or two in advance.

TO COOK THE LIONFISH HADDIES:

Place a cast-iron pan on the fire over medium-high heat. Add the olive oil; meanwhile, form half dollar one-inch thick pancakes by hand, using all of the potato mix. Place as many pancakes in the pan as it will take without overcrowding; allow them to get golden brown and delicious before flipping them over and adding the remaining butter. Baste your cakes a little while they finish cooking.

TO SERVE:

Remove them from the pan and taste and adjust seasoning if necessary. At this point they can be held at room temperature or in the oven. Garnish with lemon zest, lemon wedge, crème fraiche, mustard, the remaining dill, and additional smoked lionfish. And maybe spoon on some caviar for a classic high-low combo. Enjoy.

Chef Sawyer won the *Food & Wine* "Best New Chef" award, the James Beard Foundation Award Best Chef: Great Lakes. Based in Cleveland, Jonathan is building a small empire of restaurants including The Greenhouse Tavern, Noodlecat, and Trentina as well as a probiotic vinegar business, Tavern Vinegar Co. When Jonathon is not in the kitchen, he is hanging out with his wife Amelia, their two kids Catcher and Louisiana, their two dogs Potato and Vito, or their two backyard chickens Bear and Squid.

SWEETS

"Since we must live to eat,
we might as well do it with both grace and gusto."
—MFK Fisher

As the weather changes across the Caribbean, so does the fresh produce that is available. Each season offers an array of beautiful fresh produce; this is a great time to start experimenting and trying new recipes or reworking those old ones to incorporate more seasonal citrus and tropical fruits. In-season produce is fresher and tastes better, as it is sweeter and perfectly ripe. When fruits and vegetables are picked for consumption that have been naturally ripened on the vine or the tree and harvested at the right time, they will have much more flavor and nutrition. Most people have experienced that super juicy, perfectly sweet orange in winter, but have you ever tasted a tree-ripe summer mango?

Sweets

CARIBBEAN RUM CAKE WITH PASSION FRUIT CURD

Rum is the local go-to drink in the Caribbean. The Caribbeans are passionate about it. Each island has got its own varieties and flavors. And some of the well-aged ones are worth savoring like a good cognac. Poured over a sweet sponge and it becomes a dessert not to be taken lightly.

FOR THE RUM CAKE

- 3 eggs
- 2 cups sugar
- 10 ounces orange juice
- 12 ounces olive oil
- 3 passion fruits, pulp scooped out
- 3 teaspoons lemon zest
- 2 cups flour
- ½ teaspoon baking powder
- ½ teaspoon baking soda
- 1 teaspoon salt
- For the syrup
- 1 cup sugar
- 1 cup water
- 2 pieces star anise
- 1 vanilla pod, split in half and scraped
- 1 cup dark rum

FOR THE PASSION FRUIT CURD

- 7 ounces passion fruit puree
- 10 ounces confectionary sugar
- 12 egg yolks
- 2 passion fruits, pulp scooped out
- 9 ounces butter, cold and cubed

TO PREPARE THE RUM CAKE:

Place a pot with some water on the stove. Mix the eggs and sugar together in a heatproof bowl and place on top of the boiling pot of water. Whisk the eggs and sugar until the sugar has dissolved. Remove from the heat and place the egg mixture in a machine mixing bowl with a whisk attachment. Mix until light and fluffy. Add the orange juice, olive oil, pulp, and lemon zest. Fold in the flour, baking powder, baking soda, and salt. Pour into a one pound bread loaf pan, lined with silicone paper. Bake at 300 Fahrenheit until a skewer comes out clean and the sponge is firm and brown.

TO PREPARE THE RUM SYRUP:

In a small pan, bring the sugar and water to a boil with the star anise and vanilla pod inside. Turn down to a simmer. When the syrup starts to thicken, remove from the heat and leave to cool slightly. Then add the rum. When the sponge comes out of the oven, leave to cool a little, then pour the syrup over the top.

TO PREPARE THE PASSION FRUIT CURD:

Place a pot of water on the stove. In a heatproof bowl, mix the juice, icing sugar, pulp, and egg yolks together. Place over the boiling pot of water. Mix continuously until the mixture starts to thicken. A temperature of 185 Fahrenheit should be achieved to make sure the yolks have been cooked completely. Remove from the heat and start to mix in the butter little bits at a time. Make sure they are incorporated completely before adding the next piece of butter. Too much might make the curd split. When all the butter has been added, blend with a hand blender to finalize the emulsion. Pour into a container and cover with a piece of plastic wrap directly on top of the mixture to prevent it from forming a skin. Leave to cool overnight. Serve with the cake the next day.

MANGO MOUSSE WITH CHANTILLY CREAM AND MERINGUE

Mangoes are everywhere in summer. There are so many varieties with different flavors and textures, and it is hard to choose. They all taste amazing! Try this delicious mango mousse topped with Chantilly cream, fresh mangos, and crispy meringue.

FOR THE MANGO MOUSSE

- 1 cup mango puree
- 4 Tablespoons white sugar
- 3 leaves gelatin
- 1 cup heavy cream
- 1 large fresh mango for garnish
- For the meringue
- 1 cup egg whites
- 1 ¼ cups confectionary sugar

FOR THE CHANTILLY CREAM

- 1 cup heavy cream for whipping
- 2 Tablespoons confectionary sugar
- 1 teaspoon vanilla paste (one whole vanilla pod)

TO PREPARE THE MANGO MOUSSE:

Mix the mango puree and the white sugar together. Warm gently on the stove, but do not boil. Soak the gelatin leaves in cold ice water until soft. Add to the warm mango and sugar mixture and stir gently until the gelatin has dissolved completely. Leave to cool to room temperature on the side. Semi-whip the cream in a mixer with a whisk attachment or a handheld whisk until you're can make a ribbon or an eight shape. Fold the cream gently into the mango puree mixture. Pour into glasses, and place in the refrigerator to firm up.

TO PREPARE THE MERINGUE:

In a clean machine mixing bowl with a whisk attachment, place the egg whites and start whipping until it starts to resemble shaving foam. Slowly add the confectionary sugar teaspoons at a time until all has been added to the egg whites. When you have a firm meringue, spread or pipe this on a silicone paper or a non-stick baking sheet. Dry in an oven overnight at 165 degrees. When completely dry, place in an airtight container and reserve until necessary.

TO PREPARE THE CHANTILLY CREAM:

Place the cream, sugar, and vanilla in a machine mixing bowl with a whisk attachment. Mix until semi-whipped. Store in the refrigerator until needed.

TO ASSEMBLE:

Spoon Chantilly cream on top of the mango mousse. Peel and slice the firm ripe mango into thin slices. Place on top of Chantilly and stick the meringue into the glass.

BANANA ICE CREAM AND CHOCOLATE SAUCE

The islanders love their bananas, including green figs, as some types are locally called in savory dishes, or plantains which are great when fried for chips as a snack. But what they love most is the sweetness of fresh, ripe bananas. Here is an easy-to-make banana ice cream that doesn't require an ice cream machine with a crispy almond crumble and a warm chocolate sauce.

FOR THE ICE CREAM

- 6 medium bananas, very ripe, about to turn brown
- 1 teaspoon vanilla paste
- 2 Tablespoons light brown sugar
- 4 Tablespoons coconut cream

FOR THE ALMOND CRUMBLE

- 4 Tablespoons cold butter
- 4 Tablespoons light brown sugar
- 4 Tablespoons almond flour
- 2 Tablespoons flour
- 2 Tablespoons cocoa powder

FOR THE CHOCOLATE SAUCE

- 1 cup heavy cream
- 1 cup high-quality dark chocolate

TO PREPARE THE BANANA ICE CREAM:

Prepare the day before. Peel, cut, and freeze the ripe bananas. Leave them overnight in the freezer. When completely frozen, place the bananas in a blender or food processor. Blend until smooth in texture, as with a soft ice cream. Add the vanilla, sugar, and coconut cream as desired to taste. Blend again. Place into a container suitable for freezing. Return to the freezer for 40 minutes to an hour for a firmer ice cream texture. Scoop when ready.

TO PREPARE THE ALMOND CRUMBLE:

Preheat oven to 325 degrees. Place all the ingredients into a mixing bowl. With your hands or the paddle attachment, crumb all the ingredients together until the mix has a lightly crumbed texture with no butter lumps. Pour crumble onto a flat baking tray lined with silicone paper. Bake in the oven for 20 to 25 minutes stirring regularly until the crumble starts to smell nutty. Remove from the oven, and leave to cool.

TO PREPARE THE CHOCOLATE SAUCE:

Chop the chocolate into smaller pieces. Pour the cream in a pan. Bring to a boil, then pour over the chopped chocolate and mix together until it resembles a smooth chocolate sauce.

SPICED CHOCOLATE CHEESECAKE WITH BARBADOS CHERRY COMPOTE

Cacao has been in the Caribbean for hundreds of years, and Trinitario is a special variety originally from Trinidad. Chili gives this cheesecake a warm cozy flavor, perfect for cold winter weather. Delicate Barbados cherries have uniquely sour, but floral notes and complement this dish perfectly.

FOR THE CHEESECAKE BASE

- 2 cups graham crackers
- 4 Tablespoons butter, melted
- 2 Tablespoons quality cacao powder

FOR THE CHEESECAKE MIX

- 2 cups 70 percent dark chocolate
- ¾ cup heavy cream
- 28 ounces cream cheese
- ¾ cup brown sugar
- 1 teaspoon chili flakes (optional)
- ¾ teaspoon cinnamon powder
- 1 vanilla pod, scraped
- 4 eggs

FOR THE CHERRY COMPOTE

- 2 cups Barbados cherries (sour cherries as substitute)
- 1 cup white sugar
- 1 vanilla pod

TO PREPARE CHEESECAKE BASE:

Preheat oven to 350 degrees. Line the outside of a 10-inch cake ring or springform pan with heavy duty foil. Place a disc of silicone paper at the bottom. Then spray with non-stick spray. Blend the graham crackers in a food processor till fine, add the melted butter and cacao powder. When combined, press into cake ring. Bake for 8 to 10 min until firm and leave to cool. Reduce oven temperature to 250 Fahrenheit.

TO PREPARE THE CHEESECAKE:

Chop the chocolate until fine; bring the cream to a boil and pour it over the chocolate. Mix till melted and smooth. Separately, beat the cream cheese, brown sugar, chili flakes, and cinnamon powder in an electric mixer until smooth. Add the eggs one at a time until incorporated. Fold in the chocolate mixture. Pour over the base and bake for 1 ½ hours until the sides are starting to firm up but it still soft and wobbly in the middle. It should not form cracks. Switch oven off and leave inside for 30 to 40 minutes; it should now be firm in the middle. Remove from the oven and leave to cool. Pass a small pallet knife along the edges. Then remove the cake ring. Refrigerate for 4 hours or overnight. Place on serving plate or cake stand.

TO PREPARE THE CHERRY COMPOTE:

Remove the stones from all the cherries. Reserve some fresh cherries to garnish. Slice cherries in half. In a small pan, add the sugar, cherries, and the vanilla pod. Bring to a boil and cook till the syrup starts to thicken. Remove and leave to cool. Mix the reserved cherries, stones removed, into the cooked compote.

TO ASSEMBLE:

Pour the cherries over the cheesecake and garnish with fresh cherries.

SERVES 4

EMERALD ESTATE CHOCOLATE TRUFFLES WITH CARIBBEAN SEA SALT

Our chocolate is handcrafted with organically grown cocoa beans from Nick Troubetzkoy's Emerald Estate on Saint Lucia in the West Indies. Three generations of care and local technique have taught us to select exceptional quality cacao beans that are handpicked at their peak maturity and ripeness.

We are a bean-to-bar single estate boutique chocolate maker. I have pioneered this signature intense, aromatic flavor and deep chocolate taste. Our secret is authentically hand crafting small batches of cacao beans, slow roasted and stone ground. Emerald Estate Chocolate is to savor, but then let the exotic essence carry you to the Caribbean.

- 1 cup dark chocolate, 70 percent cocoa, chopped
- 3 Tablespoons unsalted butter, cubed
- ½ cup heavy cream
- 1½ teaspoons course Caribbean Sea salt
- 1 cup bittersweet chocolate, 60 percent cocoa, melted

TO PREPARE THE CHOCOLATE:

Place 1 cup dark chocolate and the butter in a medium heatproof bowl. In a small saucepan over medium-high heat, bring the cream and ½ teaspoon sea salt to a boil. Pour the cream over the chocolate and butter. Let stand for 2 to 3 minutes before whisking until melted and smooth. Refrigerate, uncovered, for an hour, or until slightly firmed.

TO MAKE THE TRUFFLES:

Use a small 2-teaspoon spring-loaded scoop, form mounds of chocolate on a parchment-lined baking sheet. Refrigerate 15 minutes. Roll the mounds into balls with the palms of your hands. Return the truffles to the baking sheet and chill for another 30 minutes or until firm. Dip the balls in prepared melted chocolate and return to tray.

TO FINISH THE TRUFFLES:

Let chocolate set before serving. Lightly sprinkle a spot of sea salt on the top of each individual truffle. The truffles can be made ahead of time and stored in an airtight container in the fridge for up to 3 days.

COCKTAILS

"I have never tasted anything so cool and clean.
They made me feel civilized."
—Ernest Hemmingway

Craft cocktails use fresh, in-season ingredients. Craft simply means in our corner of the Caribbean that every ingredient is crafted fresh. Fresh squeezed juices, handmade syrups, freshly pureed tropical fruits, and just-picked herbs. Basically, here are a few of our favorite seasonal and original cocktails made with craftmanship and special care for the details. The process of making craft cocktails is a lot like that of the artisan food we create throughout the resort, focusing on flavor, high-quality ingredients, and taking time to do those ingredients justice.

ANSE CHASTANET
JADE MOUNTAIN

Our Local Craft Cocktails

Non-Alcoholic Craft

MAKES 1 COCKTAIL

PINEAPPLE AND COCONUT LIME SPRITZ

This is a perfect refreshment after a day of snorkeling, watching a myriad of tropical fish, and exploring the reefs. Enjoy this cooling spritzer with your toes in the clear blue water's edge of the Caribbean.

- 2 Tablespoons fresh pineapple chunks
- 1 teaspoon lime juice
- 2 ounces coconut water
- 4 ounces prosecco

TO PREPARE COCKTAIL:

Put pineapple, lime juice, and coconut water in a shaker. Muddle all together. Add 4 to 6 solid ice cubes, cover, and shake hard. Double strain into a chilled wine glass and top with prosecco. Garnish with pineapple leaves.

MAKES 1 COCKTAIL

MANGO TI PUNCH

Saint Lucia is literally dripping with mangos, having over thirty varieties of mangoes of magnificent colors and flavors. Mango Ti Punch is short for Petit Punch which is served throughout the West Indies. Nowadays, you can find a ripe mango everywhere on the island. Mango and rum are one of the quintessential Caribbean combinations for all the right reasons: together they are lush, rich, tropical, and delicious.

- 1½ ounces Caribbean rum
- 3 ounces fresh mango puree
- 1 teaspoon lime juice
- 1 teaspoon turbinado sugar
- 2 ounces ginger ale

TO PREPARE THE COCKTAIL:

Put the rum, mango puree, lime juice, and sugar in a shaker. Cover and shake quick and hard. Strain into a large tumbler filled with ice. Top with cold ginger ale and garnish with fresh mango slice.

MAKES 1 COCKTAIL

PASSION FRUIT MOJITO

The mojito originated in Cuba but has spread as quickly as lionfish throughout the Caribbean. There is no stopping this cocktail. So, you might just want to share a few with friends. The name mojito is a slang expression meaning "a little soul." The passion fruit adds a bit of tasty rhythm to this soulful and delicious cocktail.

- ½ fresh lime cut into wedges
- 8 fresh mint leaves
- 2 Tablespoons sugar
- 1 large fresh passion fruit, flesh and seeds scooped
- 1½ oz Caribbean white rum
- 3 oz club soda

TO PREPARE THE COCKTAIL:

In a shaker, muddle the lime wedges and 4 mint leaves with sugar. Stir in the passion fruit and rum. Strain into a tall glass with 6 to 8 ice cubes and top with club soda. Garnish with remaining mint leaves.

MAKES 1 COCKTAIL

GRAPEFRUIT GRENADINE ROSE

The grapefruit is the only citrus to have originated in the Caribbean. It is a wild cross between the Asian pumelo and the local oranges. The name grapefruit come from the fact that they grow in bunches—like grapes on the tree. Pink grapefruit have yellow skin, but their fleshy interiors are pink and refreshing.

- 5 oz pink grapefruit juice
- 2 teaspoons grenadine
- 2 ounces gin
- 1 teaspoon fresh lime juice
- 1 slice fresh lime
- 1 large basil leaf

TO PREPARE THE COCKTAIL:

In a shaker, stir together the grapefruit juice, grenadine, gin, and lime juice. Pour into a tall tumbler filled with ice. Garnish with lime and basil leaf.

MAKES 1 COCKTAIL

SORREL AND GINGER TEA SMASH

This sweet, cinnamon-spiced non-alcoholic drink gets its festive crimson shade from the flowers of roselle, a species of tropical hibiscus plant used to make it. To prepare the sorrel tea, combine 2 Tablespoons of the dried hibiscus buds in a saucepan with 1 quart of water along with local spices: ½ teaspoon cinnamon, 3 whole cloves, and 1 teaspoon freshly chopped ginger. The mixture is boiled for a few minutes then steeped for 2 to 3 hours for all the flavors to blend.

- 1 thumbnail-sized fresh ginger
- 2 teaspoons guava jelly
- 8 ounces infused sorrel tea
- 2 ounces club soda
- 1 slice fresh, ripe pineapple

TO PREPARE THE COCKTAIL:
Muddle ginger with guava jelly in a shaker. Add sorrel tea and 6 to 8 ice cubes, cover and shake well. Strain into a brandy alexander glass filled with ice and top with club soda. Garnish with pineapple slice.

RESOURCES

- Monterey Bay Aquarium: www.montereybayaquarium.org
- Seafood Watch: seafoodwatch.org
- Mote Marine Laboratory & Aquarium: mote.org
- REEF—Reef Environmental Education Foundation: REEF.org
- Florida Fish and Wildlife Conservation Commission: myfwc.com
- CORE—Caribbean Oceanic Restoration & Education Foundation: corevi.org
- NOAA—National Ocean Service: Oceanservice.noaa.gov
- NOAA—National Marine Sanctuaries: Sanctuaries.noaa.gov
- Pew Charitable Trust Ocean: pewtrusts.org/oceans
- Lionfish University: lionfishuniversity.org
- CULL—Cayman United Lionfish League: cita.ky
- Barbados Lionfish Project: slowfoodbarbados.org/slow-fish
- Saint Lucia Department of Fisheries: www.govt.lc/ministries/agriculture-food-production-fisheries-and-rural-development/fisheries-department
- Whole Foods Market is a reliable source for lionfish throughout the state of Florida: wholefoodsmarket.com

SCUBAPRO

JADE MOUNTAIN

Jade Cuisine is a celebration of the bold flavors of the world's tropical cultures in my kitchen at Jade Mountain. The cooking is fresh, simple, and succinct. There is always flexibility for the seasons, and availability of local products. As a chef, my passion is taste. The bottom line here is that food comes down to nature and the nature of taste.

As you have seen visualized throughout the book, rising majestically above the six hundred-acre beach front resort of Anse Chastanet, Jade Mountain Resort on Saint Lucia's southwestern Caribbean coastline is a cornucopia of organic architecture celebrating Saint Lucia's stunning scenic beauty.

Our two Saint Lucia resorts, Anse Chastanet and Jade Mountain, were created with an environmental consciousness at heart, long before this became not only fashionable but a mandatory resort philosophy. Jade Mountain was the first resort in the Caribbean to win LEED gold. In addition, it is recognized as one of the top resorts in the world and as the best Caribbean resort, and Anse Chastanet is acknowledged as having the best Caribbean beach. Jade Mountain Club restaurant was also named one of the Top 10 Caribbean restaurants.

Architect and owner Nick Troubetzkoy has expanded upon his philosophy of building in harmony with Caribbean nature in his creation of Jade Mountain. The bold architectural design—individual bridges leading to exceptional infinity pool sanctuaries and rugged stoned-faced columns reaching toward the sky—makes Jade Mountain Saint Lucia one of the Caribbean's most mesmerizing resort experiences. With the fourth wall entirely absent, Jade Mountain's sanctuaries are stage-like settings from which to embrace the full glory of Saint Lucia's Pitons World Heritage Site and, of course, the eternal Caribbean Sea.

The Anse Chastanet reef starts just fifteen meters beyond the water's edge and can be enjoyed as a shore dive. The reef's remarkable ecosystem offers an amazing profusion of unusual tropical marine life in 20 to 140 feet of calm, clear water. The Anse Chastanet Reef, which is home to more than 150 different species of fish, makes for an amazing dive, day or night. In the shallow areas, be sure to keep your eyes open for peacock flounders, octopus, needlefish, and turtles or drop down a little deeper over dense coral growth to see puffers, moray eels, parrot fish, lobsters, sea horses, and even lionfish. The resort's team manages the protected near shore area. In fact, the resorts played an instrumental role in the creation and implementation of the Soufriere Marine Management Authority, which has led to the entire near shore marine environment to be declared a marine reserve.

ACKNOWLEDGMENTS

All my thanks and admiration go to my friends and team at Jade Mountain, Anse Chastanet, Emerald Estate, and Scuba St Lucia, especially owners Nick and Karolin Troubetzkoy who have been my inspiration and a part of my Caribbean experience. Their creative drive, infused with a concordant sense of nature and a mantra of sustainability, has allowed me to develop Jade Cuisine with a sense of place.

Karolin, this book had its beginnings with you, when you asked me if there was something we could do with lionfish. That request evolved into a weekly beachfront Friday night lionfish tasting experience. And recipes followed. Realizing how lionfish were negatively impacting the Caribbean, you encouraged me to write this book to share with guests, divers, and everyone who loves the natural beauty of the Caribbean. You have great intuition.

Thank you, Mr. Nick, for your vision and dynamic thinking. I appreciate your constant challenge to think big and to create social value. And for providing me a space on your most natural canvas, allowing me to share my creativity.

Thank you, Andreas Nangele and Yasha Troubetzkoy, who both appreciate food as a way of life.

To Chef Elijah, who is keeping it real. You have shared your Saint Lucian way of life and cooking with me for many years. I appreciate your sense of cooking and great talent for making Jade Cuisine taste as good as it looks. You have the eye for plating food.

I also want to show my appreciation for the entire culinary team at the resorts whose help continues to be vibrant, including the vital leadership of Exec Chef Juan, Chef Adam, Chef Frankie, and Chef Ruban.

Thank you to our chocolate team from tree to bar to pleasure. Including the Emerald Estate team and Farm Manager Matt, Chocolate Lab Chocolatier Peter, and Exec Pastry Chef Alexander.

My thanks go out to Bernd Rac and his camera. Just about every beautiful photo in this book is shot by Bernd. I appreciate your patience and photographic gift.

Thank you, Scuba St Lucia team, led by Jay and Gigi, for your insight into the world of lionfish. In addition to the stunning underwater action shots.

I would like to thank my friends at Mango Publishing and Books & Books Press, Mitchell Kaplan and Chris McKenney, for giving me the opportunity to do this book.

Thank you to everyone at Mango for their support and encouragement, especially to Morgane Leoni for the distinctive layout and design of this book. And to my talented marketing team, Hannah Jorstad Paulsen and Merritt Smail, who empowered me to reach out for my vision. Much appreciation goes out to my insightful editor, Yaddyra Peralta, for her creative skills and deep thinking on the craft which helped me to organize my writing into English others could understand.

ABOUT THE AUTHOR

Chef Allen Susser is a James Beard award-winning chef. He has a passionate commitment to local fresh ingredients. *The New York Times* called Allen the "Ponce De Leon of New Florida cooking." His landmark restaurant Chef Allen's forever changed the way people ate in Miami and affected how America eats today. *Food & Wine* magazine named Chef Allen "One of the Best 10 Chefs in America."

Chef Allen's Consulting is a boutique sustainable restaurant and hospitality consulting firm, providing strategic initiative, culinary resources, and innovative direction for the industry. Allen is the consulting chef to Jade Mountain and Anse Chastanet, helping to create one of the most dynamic culinary destinations in the Caribbean. This includes pioneering Emerald Estate Vintage Chocolate, a bean-to-bar authentically handcrafted organic chocolate from Saint Lucia in the West Indies.

Susser is currently a Culinary Ambassador for the Monterey Bay Aquarium for its sustainable seafood watch. Allen recently was awarded a lifetime achievement award for his farm-to-table community involvement with Slow Foods. He received an honorary Doctor of Culinary Arts from Johnson and Wales University, as well as the President's Award for community service from Florida International University. He was recognized as the humanitarian of the year by Share Our Strength for his commitment to End Childhood Hunger. Allen is the author of *New World Cuisine and Cookery*, *The Great Citrus Book*, and *The Great Mango Book*.

Mango Publishing, established in 2014, publishes an eclectic list of books by diverse authors—both new and established voices—on topics ranging from business, personal growth, women's empowerment, LGBTQ studies, health, and spirituality to history, popular culture, time management, decluttering, lifestyle, mental wellness, aging, and sustainable living. We were recently named 2019's #1 fastest growing independent publisher by *Publishers Weekly.* Our success is driven by our main goal, which is to publish high quality books that will entertain readers as well as make a positive difference in their lives.

Our readers are our most important resource; we value your input, suggestions, and ideas. We'd love to hear from you—after all, we are publishing books for you!

Please stay in touch with us and follow us at:

Facebook: Mango Publishing
Twitter: @MangoPublishing
Instagram: @MangoPublishing
LinkedIn: Mango Publishing
Pinterest: Mango Publishing

Sign up for our newsletter at **www.mango.bz** and receive a free book!

Join us on Mango's journey to reinvent publishing, one book at a time.